STU̶ ̶ ̶ ̶ ̶ ̶ ̶ ̶ ̶UAL

Probability and Statistics for Engineers

FOURTH EDITION

Richard L. Scheaffer
University of Florida

James T. McClave
Infotech, Inc.

Prepared by
Chris Franklin
University of Georgia

Duxbury Press
An Imprint of Wadsworth Publishing Company
Belmont, California

International Thomson Publishing
The trademark ITP is used under license.

Duxbury Press
An Imprint of Wadsworth Publishing Company
A Division of Wadsworth, Inc.

© 1995 by Wadsworth, Inc. All rights reserved. No part of this book may be reproduced, stored in a retrieval system, or transcribed, in any form or by any means, without the prior written permission of the publisher, Wadsworth Publishing Company, Belmont, California 94002.

Printer: Malloy Lithographing, Inc.

Printed in the United States of America

2 3 4 5 6 7 8 9 10—04 03 02 01 00

ISBN 0-534-20965-3

Contents

Acknowledgments

I would like to express my sincere appreciation to Michael Allen and Deborah Allen for their assistance in the preparation of this manual. And to Connie Mobley, my typist.

Chris Franklin

Chapter 1.

Minitab output for selected problems is contained in the Appendix. Numerical answers given below were obtained using Minitab.

1.7

    b. $\bar{x} = 10.35$ $s = 49.00$

    d. No. The extreme data point at $+205.2$ inflates the mean and standard deviation.

1.11   Yes.

1.13

    c. $\bar{x}_1 = 4.37$ $s_1 = 0.34$

       $\bar{x}_2 = 1.12$ $s_2 = 0.82$

1.15 $\bar{x}_1 = 83.4$ $s_1 = 89.20$

       $\bar{x}_2 = 35.8$ $s_2 = 39.16$

1.17   SAT: $\bar{x} = 474.30$ $s = 6.96$

       ACT: $\bar{x} = 17.75$ $s = 0.84$

    f. 18.7

1.19 $\bar{x} = 0.344$ $s = 0.022$.

    c. Yes.

    d. Yes.

Chapter 2.

2.1   Discussion answers. Part (c) is definitely a subjective statement since there is no scientific basis to prove or disprove the probability.

2.3   Discussion answer.

2.5   Discussion answer.

2.7   Discussion answer.

2.9

    a. 33%

    b. Observational study.

2.11 There are 10 elements in P, 5 in M, and 13 in neither P nor M. Since there are only 25 elements in all, $P \cap M$ must contain $28 - 25 = 3$ elements. This setting is described by the following Venn diagram.

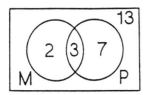

    a. PM contains 3 elements.

    b. $\overline{PM}$ contains 13 elements.

    c. $P\overline{M}$ contains 7 elements.

    d. $(\overline{PM} \cup \overline{P}M)$ contains $2 + 7 = 9$ elements.

2.13 Let J denote the event that Jim gets one of the jobs and D denote the event that Don gets one of the jobs; similarly define M, S, and N for Mary, Sue, and Nancy, respectively.

a. Considering unordered outcomes (i.e., JD is the same outcome as DJ), all possible selections of two applicants from the five are:

JD	JM
JS	JN
DM	DS
DN	MS
MN	SN

b. $A = \{JD, \quad JM, \quad JS, \quad JN, \quad DM, \quad DS, \quad DN\}$; i.e., there are 7 elements in A.

c. $B = \{JM, \quad JS, \quad JN, \quad DM, \quad DS, \quad DN\}$; i.e., there are 6 elements in B.

d. The set containing two females is the complement of the set containing at least one male; i.e., $\overline{A} = \{MS, \quad MN, \quad SN\}$.

e. $\overline{A} = \{MS, \quad MN, \quad SN\}$

$AB = B = \{JM, \quad JS, \quad JN, \quad DM, \quad DS, \quad DN\}$

$A \cup B = A = \{JD, \quad JM, \quad JS, \quad JN, \quad DM, \quad DS, \quad DN\}$

$\overline{AB} = \{JD, \quad MS, \quad MN, \quad SN\}$

2.15 Verification of $\overline{AB} = \overline{A} \cup \overline{B}$:

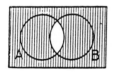

IIII for $\overline{AB}$

Shaded region is $\overline{AB}$

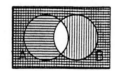

IIII for $\overline{A}$,

for $\overline{B}$

Region shaded at least once is $\overline{A} \cup \overline{B}$

3

Verification of $\overline{A \cup B} = (\overline{A})(\overline{B})$:

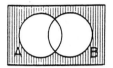
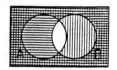

IIII for $\overline{A \cup B}$

Shaded region is $\overline{A \cup B}$

III for $\overline{A}$,

Crosshatched region is $(\overline{A})(\overline{B})$

for $\overline{B}$

2.17

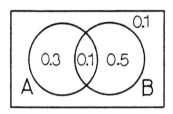

Using the above Venn diagram, we have the following:

a. $P(A) = 0.3 + 0.1 = 0.4$

b. $P(A \cup B) = 0.9$

c. $P(\overline{B}) = 0.4$

d. $P(AB) = 0.1$

e. $P(A \cup \overline{B})\ 0.5$

f. $P(\overline{A} \cap \overline{B}) = 0.1$

g. $P(\overline{A \cup B}) = 0.1$

2.19

a. 0.43

b. 0.05

c. $0.43 + 0.12 + 0.04 = 0.59$

d. $0.43 + 0.31 = 0.74$

e. $0.43 + 0.31 + 0.05 + 0.12 = 0.91$

2.21 Denote an outcome as an ordered pair of letters, where the first letter of the pair signifies the outcome for the first vehicle (and may assume the values S = straight, L = left turn, R = right turn) and the second letter signifies the outcome for the second vehicle (and may similarly assume the values S, L, or R).

a.

Vehicle I	Vehicle II	Outcome
Straight	Straight	SS
Straight	Right	SR
Straight	Left	SL
Right	Straight	RS
Right	Right	RR
Right	Left	RL
Left	Straight	LS
Left	Right	LR
Left	Left	LL

b. P(at least one turns left) = P(SL) + P(RL) + P(LS) + P(LR) + P(LL)

$$= \frac{1}{9} + \frac{1}{9} + \frac{1}{9} + \frac{1}{9} + \frac{1}{9} = \frac{5}{9}$$

c. P(at most, one vehicle turns) = P(SS) + P(SR) + P(SL) + P(RS) + P(LS)

$$= \frac{1}{9} + \frac{1}{9} + \frac{1}{9} + \frac{1}{9} + \frac{1}{9} = \frac{5}{9}$$

2.23 Denote an outcome as an ordered pair of Roman numerals, where the first Roman numeral of the pair signifies the firm (either I, II, or III) that receives the first contract and the second signifies the firm that receives the second contract.

a. The simple events are (I, I), (I, II), (I, III), (II, I), (II, II) (II, III), (III, I), (III, II), (III, III).

b. P(both contracts go to same firm) = P[(I, I)] + P[(II, II)] + P[(III, III)]

$$= \frac{1}{9} + \frac{1}{9} + \frac{1}{9} = \frac{1}{3}$$

c. P(firm I gets at least one contract) = P[(I, I)] + P[(I, II)] + P[(I, III)]

$$+ \text{P}[(\text{II, I})] + \text{P}[(\text{III, I})]$$

$$= \frac{1}{9} + \frac{1}{9} + \frac{1}{9} + \frac{1}{9} + \frac{1}{9} = \frac{5}{9}$$

2.25

    a. $P_2^7 = 7(6) = 42$

    b. $\binom{7}{2} = \frac{7(6)}{2(1)} = 21$

2.27 $P_4^{10} = 10 \cdot 9 \cdot 8 \cdot 7 = 5{,}040$

2.29 $\dbinom{n-1}{r-1} + \dbinom{n-1}{r} = \dfrac{(n-1)!}{(r-1)!(n-r)!} + \dfrac{(n-1)!}{r!(n-1-r)!}$

$$= \frac{r(n-1)!}{r(r-1)!(n-r)!} + \frac{(n-r)(n-1)!}{(n-r)r!(n-1-r)!}$$

$$= \frac{r(n-1)! + (n-r)(n-1)!}{r!(n-r)!}$$

$$= \frac{n!}{r!(n-r)!} = \binom{n}{r}$$

2.31

    a. $P\begin{pmatrix} \text{one blue, one white,} \\ \text{and one green ordered} \end{pmatrix} = \dfrac{\begin{pmatrix} \text{number of ways one blue,} \\ \text{one white, and one green} \\ \text{can be ordered} \end{pmatrix}}{\begin{pmatrix} \text{total number of ways} \\ \text{a sample of three} \\ \text{can be chosen} \end{pmatrix}}$

$$= \frac{3!}{4^3} = \frac{6}{64} = \frac{3}{32} = 0.09375$$

b. $P\left(\begin{array}{c} \text{two blues} \\ \text{are ordered} \end{array}\right)$

$$= \frac{\left(\begin{array}{c} \text{number of ways of} \\ \text{ordering two blues} \\ \text{out of an order of three} \end{array}\right)\left(\begin{array}{c} \text{number of ways (colors)} \\ \text{the remaining car} \\ \text{can be ordered} \end{array}\right)}{\left(\begin{array}{c} \text{total number of ways} \\ \text{a sample of three} \\ \text{can be chosen} \end{array}\right)}$$

$$= \frac{\binom{3}{2}\binom{3}{1}}{4^3} = \frac{9}{64} = 0.140625$$

c. Note that if no black are chosen, then each order may be any of the three remaining colors, and this may be done in $3^3 = 27$ ways. Thus P(at least one black) $= 1 - $P(none are black) $= 1 - \dfrac{3^3}{4^3} = 1 - \dfrac{27}{64} = \dfrac{37}{64} = 0.578125$.

d. Multiplying the answer to (b) by the number of ways to choose a duplicated color, we have

$$\frac{\binom{4}{1}\binom{3}{2}\binom{3}{1}}{4^3} = \frac{36}{64} = 0.5625.$$

2.33

a. $P_4^4 = 4! = 24$

b. Required probability

$$= \frac{\left(\begin{array}{c} \text{number of orderings} \\ \text{in event of interest} \end{array}\right)\left(\begin{array}{c} \text{number of ways the remaining three} \\ \text{operations may be performed} \end{array}\right)}{\left(\begin{array}{c} \text{total number of ways the} \\ \text{operation can be performed} \end{array}\right)}$$

c. $\qquad = \dfrac{2 \cdot 3!}{4!} = \dfrac{1}{2}$

2.35 The sample space is {SS, SR, SL, RS, RR, RL, LS, LR, LL}. Let A be the event of at least one vehicle turning left and B be the event that at least one vehicle turns. Assuming equally likely outcomes, we have

$$P(A|B) = \frac{P(AB)}{P(B)} = \frac{P(A)}{P(B)}$$

$$= \frac{P(SL \cup RL \cup LS \cup LR \cup LL)}{1 - P(SS)} = \frac{5/9}{8/9} = \frac{5}{8}.$$

2.37

a. $\dfrac{46,263}{92,911} = 0.4979$

b. Let V = motor vehicle accident, M = male.

$$P(V|M) = \frac{P(VM)}{P(M)} = \frac{32,949}{64,053} = 0.5144$$

c. Let Y = the event of $(15 \le \text{age} \le 24)$. Then

$$P(V|Y) = \frac{P(VY)}{P(Y)} = \frac{14,738}{19,801} = 0.7443$$

d. Let F = event of accidental death by falling, and E = event of age $\ge$ 75. Then

$$P(F|E) = \frac{P(FE)}{P(E)} = \frac{7,067}{16,065} = 0.4399$$

e. $P(M) = \dfrac{64,053}{92,911} = 0.6894$

2.39

a. 0.10

b. 0.03

c. 0.06

d. Let M= mobile home fire, S = fire caused by smoking. Then

$$P(M|S) = \frac{P(MS)}{P(S)} = \frac{P(S|M)P(M)}{P(S)} = \frac{0.06(0.03)}{0.10} = 0.018$$

2.41

    a.

$$\frac{\dbinom{18}{2}}{\dbinom{20}{2}} = \frac{18 \cdot 17}{20 \cdot 19} = 0.8053$$

    b. P(at least one is nondefective) $= 1 - $ P(both are defective)

$$= 1 - \frac{\dbinom{2}{2}}{\dbinom{20}{2}} = 1 - \frac{2}{20 \cdot 19} = \frac{378}{380} = 0.9947$$

    c. $P(A|B) = \dfrac{P(AB)}{P(B)} = \dfrac{P(A)}{P(B)} = \dfrac{0.8053}{0.9947} = 0.8096$

2.43

    a. Let A_i = event that i^{th} resistor has resistance between 9.5 and 10.5 ohms,

       $i = 1, 2$. Then

       $P(A_i) = 1 - 0.05 - 0.10 = 0.85$, and

       P(both have actual values between 9.5 and 10.5)

       $= P(A_1 A_2) = P(A_1)P(A_2) = (0.85)(0.85) = 0.7225.$

    b. Let E_i = event that i^{th} resistor has resistance in excess of 10.5 ohms,

       $i = 1, 2$. Then P(at least one has an actual value greater than 10.5)

       $= P(E_1 E_2 \cup E_1 \overline{E}_2 \cup \overline{E}_1 E_2) = P(E_1)P(E_2) + P(E_1)P(\overline{E}_2) + P(\overline{E}_1)P(E_2)$

       $= (0.1)(0.1) + 2(0.1)(0.9) = 0.19 \; or$

       P(at least one has an actual value greater than 10.5)

       $= 1 - P(\overline{E}_1 \overline{E}_2) = 1 - (0.9)^2 = 0.19.$

2.45   Let C_i = event that relay i closes properly, $i = 1, 2$. Then

        P(current flows in series system)

         $= P$(both relays are closed)

          $= P(C_1 C_2) = P(C_1)P(C_2) = (0.9)(0.9) = 0.81$

P(current flows in parallel system)

$= P$(at least one of the relays is closed)

$= P(C_1 \cup C_2) = P(C_1) + P(C_2) - P(C_1 C_2)$

$= 0.9 + 0.9 - (0.9)(0.9) = 0.99$

2.47   Let F denote the event that a worker fails to learn the skill correctly.

$$P(A|F) = \frac{P(A)P(F|A)}{P(F|A)P(A) + P(F|B)P(B)} = \frac{(0.70)(0.20)}{(0.20)(0.70) + (0.10)(0.30)}$$

$= 0.8235$

2.49

   a. 4.7% of the purchases made for Aerobic shoes are for users under 14 years of age.

   b. P(the next purchase of a jogging/running shoe will be by a 25−34 year old) $= 28.5$.

   c. 34,600 for age under 14 and 9,700 by 18−24 year olds.

   d. No.

   e. By gender, more males purchase jogging/running shoes while more females purchase aerobic shoes. By age, the most purchases for sports footwear are by the age group 25−64 years old.

2.51

   a. The percentages in the bar graph for the black population are only for the black population and cannot be used to give the percentage of, for example, Hispanic people in the midwest.

   b. No, that is not a correct interpretation. One should interpret the graph as expressing that more than half of the black population lives in the south.

   c. There is a definite correlation between race and region.

2.53

a.

	Yes	No	Total
White defendant	19	141	160
Black defendant	17	149	166
Total	36	290	426

Odds Ratio:

$$\left(\frac{19}{141}\right) / \left(\frac{17}{149}\right) = 1.18$$

b.

White victim	Yes	No	Total
White defendant	19	132	151
Black defendant	11	52	63
Total	30	184	214

Odds Ratio:

$$\left(\frac{19}{132}\right) / \left(\frac{11}{52}\right) = 0.68$$

Black victim	Yes	No	Total
White defendant	0	9	9
Black defendant	6	97	103
Total	6	106	112

Odds Ratio:

$$\left(\frac{0}{9}\right) / \left(\frac{6}{97}\right) = 0$$

c. Part (b) does not imply part (a). The apparent paradox is that we are trying to use a percentage based on one category on all categories.

2.55   Let the outcome of the four tosses be denoted by an ordered quartet of letters, in which the ith letter of the quartet is H if the ith toss resulted in heads, or T if the ith toss resulted in tails, $i = 1, 2, 3, 4$.

a. The sixteen outcomes for the experiment are

HHHH	THHH
HHHT	THHT
HHTH	THTH
HHTT	THTT
HTHH	TTHH
HTHT	TTHT
HTTH	TTTH
HTTT	TTTT

b. A = {HHHT, HHTH, HTHH, THHH}

c. Assuming that all outcomes are equally likely, the probability of each outcome is 1/16 and

$$P(A) = 4 \cdot \frac{1}{16} = \frac{1}{4}$$

2.57

a. Let C_1 = event of a car traveling at less than 55 mph on a rural interstate.

Year	$P(C_1)$
1973	$0.05 + 0.02 = 0.07$
1974	$0.24 + 0.07 + 0.01 = 0.32$
1975	$0.23 + 0.05 + 0.01 = 0.29$

b. Let T = event of a truck traveling at less than 55 mph on a rural interstate.

Year	$P(T)$
1973	$0.15 + 0.05 + 0.02 + 0.01 = 0.23$
1974	$0.29 + 0.11 + 0.02 + 0.01 = 0.43$
1975	$0.29 + 0.08 + 0.02 = 0.39$

c. Let C_2 = event of a car traveling at less than 55 mph on a rural secondary road.

Year	$P(C_2)$
1973	$0.19 + 0.19 + 0.11 + 0.06 + 0.04 = 0.59$
1974	$0.23 + 0.25 + 0.14 + 0.09 + 0.04 = 0.75$
1975	$0.26 + 0.21 + 0.10 + 0.06 + 0.02 = 0.65$

2.59 $\binom{3}{0} + \binom{3}{1} + \binom{3}{2} + \binom{3}{3} = 1 + 3 + 3 + 1$

$$= 8 = 2^3$$

Recall that $(a+b)^n = \sum_{i=0}^{n} \binom{n}{i} a^i b^{n-i}$

Therefore, $(1+1)^n = \sum_{i=0}^{n} \binom{n}{i} 1^i 1^{n-i}$; i.e., $2^n = \sum_{i=0}^{n} \binom{n}{i}$

2.61 $\binom{10}{3} = \dfrac{10!}{7!3!} = \dfrac{10 \cdot 9 \cdot 8}{3 \cdot 2} = 120$

2.63 $P_3^{10} = 10 \cdot 9 \cdot 8 = 720$

2.65 $P_8^8 = 8! = 40,320$

2.67 Denote success on try i as S_i, failure as $F_i, i = 1, \quad 2, \quad 3$.

a. $P(S_1 S_2 S_3) = P(S_1)P(S_2)P(S_3) = (0.6)^3 = 0.216$

b. $P(\text{at least one success}) = 1 - P(\text{no successes}) = 1 - P(F_1 F_2 F_3)$

$$= 1 - P(F_1)P(F_2)P(F_3) = 1 - (0.4)^3 = 0.936$$

c. $P(\text{at least two successes}) = P(\text{two successes}) + P(\text{three successes})$

$$= P(S_1 S_2 F_3 \cup S_1 F_2 S_3 \cup F_1 S_2 S_3) + P(\text{three successes})$$

$$= P(S_1 S_2 F_3) + P(S_1 F_2 S_3) + P(F_1 S_2 S_3) + P(\text{three successes})$$

$$= 3(0.6)(0.6)(0.4) + (0.6)^3 = 0.648.$$

2.69   Using the notation and results of Exercise 2.68, we have

P(match occurs for first time on fourth toss) $= P(\overline{M}_1\overline{M}_2\overline{M}_3 M_4)$

$= P(\overline{M}_1)P(\overline{M}_2)P(\overline{M}_3)P(M_4)$

$= [1 - P(M_i)]^3 P(M_i) = \left(1 - \dfrac{1}{2}\right)^3 \dfrac{1}{2} = \left(\dfrac{1}{2}\right)^4 = \dfrac{1}{16}.$

2.71   Denote a win for gambler Jones on toss i as W_i, and a loss as L_i, $i = 1, 2, \ldots$.
Then $P(W_i) = P(L_i) = \dfrac{1}{2}.$

  a. P(break even after 6 tosses) = P(3 losses and 3 wins in 6 tosses)

$= \left(\begin{array}{c}\text{number of ways of}\\ \text{drawing 3 from 6}\end{array}\right) P\left(\begin{array}{c}\text{throw 3 wins and 3 losses}\\ \text{in a given sequence}\end{array}\right) = \dbinom{6}{3}\left(\dfrac{1}{2}\right)^6$

$= 20\left(\dfrac{1}{64}\right) = \dfrac{5}{16}$

  b. P(Jones wins on 10th toss)

$= P\left\{\begin{array}{c}[(5 \text{ wins and 1 loss}) \\ \vdash 1\text{st } 6 \text{ tosses} \dashv\end{array}\cap\begin{array}{c}(1 \text{ win and 1 loss}) \\ \vdash \text{next } 2 \text{ tosses} \dashv\end{array}\cap\begin{array}{c}(2 \text{ wins})] \\ \vdash 9\text{th and 10th tosses} \dashv\end{array}\right.$

$\left.\cap\begin{array}{c}[(4 \text{ wins and 2 losses}) \\ \vdash 1\text{st } 6 \text{ tosses} \dashv\end{array}\cap\begin{array}{c}(4 \text{ wins})] \\ \vdash \text{next } 4 \text{ tosses} \dashv\end{array}\right\}$

$= \left[\dbinom{6}{5}\left(\dfrac{1}{2}\right)^6 \cdot \dbinom{2}{1}\left(\dfrac{1}{2}\right)^2 \cdot \dfrac{1}{2}\cdot\dfrac{1}{2}\right] + \left[\dbinom{6}{4}\left(\dfrac{1}{2}\right)^6 \cdot \left(\dfrac{1}{2}\right)^4\right]$

$= \left(\dfrac{1}{2}\right)^{10}(12 + 15) = 27\left(\dfrac{1}{2}\right)^{10}$

2.73 $P(A) > 0$ and $P(B) > 0$, but $P(AB) = 0 \neq P(A)P(B)$ since A and B are mutually
exclusive. Therefore, A and B are not independent.

2.75 Denote the event of a defective fan by D. We have $P(A) = 0.9, P(B) = 0.1, P(AB) = 0, P(D|A) = 0.05$, and $P(D|B) = 0.03$. Then

$$P(B|D) = \frac{P(D|B)P(B)}{P(D|B)P(B) + P(D|A)P(A)}$$

$$= \frac{(0.03)(0.1)}{(0.03)(0.1) + (0.05)(0.9)} = \frac{1}{16} = 0.0625.$$

2.77 $P(A\overline{B}) = P(A) - P(AB) = P(A) - P(A)P(B)$ (because of independence)

$$= P(A)[1 - P(B)] = P(A)P(\overline{B})$$

2.79 In order to form a triangle from the line segments ax, bx, and ac we must have the following:

$$ax + ac > bx \text{ for } a < x < c$$
$$bx + ac > ax \text{ for } c < x < b$$

i.e., for $a < x < c$:

$$(x - a) + \frac{b-a}{2} > (b - x)$$

i.e., $x > \dfrac{1}{2}\left(b + a - \dfrac{b-a}{2}\right)$

i.e., $x > c - \dfrac{b-a}{4}$

and, for $c < x < b$:

$$(b - x) + \frac{b-a}{2} > (x - a)$$

i.e., $x < \dfrac{1}{2}\left(b + a - \dfrac{b-a}{2}\right)$

i.e., $x < c - \dfrac{b-a}{4}.$

Therefore, we must have

$$\left(c - \frac{b-a}{4}\right) < x < \left(c + \frac{b-a}{4}\right);$$

i.e., x must fall within a segment half the length ab. Therefore, the desired probability is $\frac{1}{2}$.

2.81 $N-1$ physical dividers are required to divide n balls into N groups (cells or boxes), $n \geq N$. Therefore, the number of ways of arranging n balls into N cells is equal to the number of ways of choosing n units out of $(N + n - 1)$ units (the n chosen units are declared to be "balls" and the remaining $N - 1$ units are declared to be "dividers"); i.e., there are

$$\binom{N+n-1}{n} = \binom{N+n-1}{N-1}$$

ways of allocating n balls to N cells.

If no cell is empty, then we have $n - N$ balls whose placement is unspecified, and following the above reasoning, there are

$$\binom{(n-N)+N-1}{n-N} = \binom{n-1}{N-1}$$

ways of allocating these $n - N$ balls to N cells.

Therefore,

$$P\left(\begin{array}{c} \text{no box will} \\ \text{be empty} \end{array}\right) = \frac{\left(\begin{array}{c} \text{number of ways of allocating} \\ n-N \quad \text{balls to} \quad N \quad \text{cells} \end{array}\right)}{\left(\begin{array}{c} \text{number of ways of allocating} \\ n \quad \text{balls to} \quad N\text{cells} \end{array}\right)}$$

$$= \frac{\binom{n-1}{N-1}}{\binom{N+n-1}{N-1}}.$$

2.83   Denote the event of a person having hepatitis as H and a positive test result as T.

We are given $P(T|H) = 0.9, P(T|\overline{H}) = 0.01$, and $P(H) = \dfrac{1}{10,000}$.

a. $P(H|T) = \dfrac{P(T|H)P(H)}{P(T|H)P(H) + P(T|\overline{H})P(\overline{H})} = \dfrac{(0.9)(0.0001)}{(0.9)(0.0001) + (0.01)(0.9999)}$

$= 0.00892$

b. Here, $P(H) = \dfrac{1}{2}$.

$$P(H|T) = \dfrac{(0.9)(0.5)}{(0.9)(0.5) + (0.01)(0.5)} = 0.9890$$

2.85 $P(A \cup B \cup C) = P[A \cup (B \cup C)]$

$= P(A) + P(B \cup C) - P[A \cap (B \cup C)]$

$= P(A) + P(B) + P(C) - P(BC) - P[A \cap (B \cup C)]$

But $P[A \cap (B \cup C)] = P(AB \cup AC) = P(AB) + P(AC) - P(ABC)$.

Hence, $P(A \cup B \cup C) = P(A) + P(B) + P(C) - P(AB) - P(AC) - P(BC)$

$+ P(ABC)$.

2.87   Assuming that each of the 365 days in the year is equally likely to be a birthday, we have the following:

P(at least two students have the same birthday)

$= 1 - $ P(none have the same birthday)

$= 1 - \dfrac{\left(\begin{array}{c}\text{number of ways none have} \\ \text{the same birthday}\end{array}\right)}{\left(\begin{array}{c}\text{total number of ways 23} \\ \text{birthdays can be chosen}\end{array}\right)}$

$= 1 - \dfrac{\left(\begin{array}{c}\text{number of ways of sampling 23} \\ \text{birthdays without replacement}\end{array}\right)}{\left(\begin{array}{c}\text{number of ways of sampling 23} \\ \text{birthdays with replacement}\end{array}\right)}$

$= 1 - \dfrac{P_{23}^{365}}{(365)^{23}} = 1 - 0.4927 = 0.5073$

Chapter 3.

3.1 $P(X = 0) = P(3 \text{ males chosen})$

$$= \frac{\left(\begin{array}{c} \text{number of ways of} \\ \text{choosing 3 out of 4} \end{array} \right) \left(\begin{array}{c} \text{number of ways of} \\ \text{choosing 0 out of 6} \end{array} \right)}{\left(\begin{array}{c} \text{total number of ways} \\ \text{of choosing 3 out of 10} \end{array} \right)} = \frac{\binom{4}{3} \binom{6}{0}}{\binom{10}{3}} = \frac{1}{30}$$

$$P(X = 1) = P(2 \text{ males and 1 female chosen}) = \frac{\binom{4}{2} \binom{6}{1}}{\binom{10}{3}} = \frac{3}{10}$$

$$P(X = 2) = \frac{\binom{4}{1} \binom{6}{2}}{\binom{10}{3}} = \frac{1}{2}$$

$$P(X = 3) = \frac{\binom{4}{0} \binom{6}{3}}{\binom{10}{3}} = \frac{1}{6}$$

3.3 $P(X = 0) = \binom{3}{0} (0.363)^0 (0.637)^3 = 0.2585$

$P(X = 1) = \binom{3}{1} (0.363)^1 (0.637)^2 = 0.4419$

$P(X = 2) = \binom{3}{2} (0.363)^2 (0.637)^1 = 0.2518$

$P(X = 3) = \binom{3}{3} (0.363)^3 (0.637)^0 = 0.0478$

This answer assumes independence of up-at-bats. This assumption may not be reasonable, since pitchers may change between up-at-bats. Boggs might get tired as the game progresses, etc. It appears that it is not unusual for a good hitter to go 0 for 3 in one game, since the probability of this for Boggs is more than $\frac{1}{4}$.

3.5   The probability that a given residential fire is in a family home is 0.73.

   a.

x	$p(x)$	
0	$\binom{4}{0}(0.27)^4$	$= 0.0053$
1	$\binom{4}{1}(0.27)^3(0.73)$	$= 0.0575$
2	$\binom{4}{2}(0.27)^2(0.73)^2$	$= 0.2331$
3	$\binom{4}{3}(0.27)(0.73)^3$	$= 0.4201$
4	$\binom{4}{4}(0.73)^4$	$= 0.2840$

   b. $P(x \geq 1) = 1 - P(X = 0) = 1 - 0.0053 = 0.9947$

3.7 $p(x) = \binom{3}{x}\left(\frac{1}{3}\right)^x\left(\frac{2}{3}\right)^{3-x}$; $x = 0, \quad 1, \quad 2, \quad 3$

   $p(y) = \binom{3}{y}\left(\frac{1}{15}\right)^y\left(\frac{14}{15}\right)^{3-y}$; $y = 0, \quad 1, \quad 2, \quad 3$

x	$p(x)$	y	$p(y)$
0	8/27	0	2,744/3,375
1	12/27	1	588/3,375
2	6/27	2	42/3,375
3	1/27	3	1/3,375

   $P(X + Y = 0) = P(X = 0)P(Y = 0)$

   $= \dfrac{8}{27} \cdot \dfrac{2744}{3375} = 0.24090$

$$P(X + Y = 1) = P(X = 0)P(Y = 1) + P(X = 1)P(Y = 0)$$

$$= \frac{8}{27} \cdot \frac{588}{3375} + \frac{12}{27} \cdot \frac{2744}{3375} = 0.41297$$

$$P(X + Y = 2) = P(X = 1)P(Y = 1) + P(X = 2)P(Y = 0)$$
$$+ P(Y = 2)P(X = 0)$$

$$= \frac{12}{27} \cdot \frac{588}{3375} + \frac{6}{27} \cdot \frac{2744}{3375} + \frac{8}{27} \cdot \frac{42}{3375} = 0.26179$$

$$P(X + Y = 3) = P(X = 0)P(Y = 3) + P(X = 1)P(Y = 2)$$
$$+ P(X = 2)P(Y = 1) + P(X = 3)P(Y = 0)$$

$$= \frac{8}{27} \cdot \frac{1}{3375} + \frac{12}{27} \cdot \frac{42}{3375} + \frac{6}{27} \cdot \frac{588}{3375} + \frac{1}{27} + \frac{2744}{3375} = 0.07445$$

$$P(X + Y = 4) = P(X = 1)P(Y = 3) + P(X = 2)P(Y = 2)$$
$$+ P(X = 3)P(Y = 1)$$

$$= \frac{12}{27} \cdot \frac{1}{3375} + \frac{6}{27} \cdot \frac{42}{3375} + \frac{1}{27} \cdot \frac{588}{3375} = 0.00935$$

$$P(X + Y = 5) = P(X = 2)P(Y = 3) + P(X = 3)P(Y = 2)$$

$$= \frac{6}{27} \cdot \frac{1}{3375} + \frac{1}{27} \cdot \frac{42}{3375} = 0.00053$$

$$P(X + Y = 6) = P(X = 3)P(Y = 3) = \frac{1}{27} \cdot \frac{1}{3375} = 0.00001$$

3.9

a.

$$p(x) = \frac{\left(\begin{array}{c}\text{number of ways of} \\ \text{choosing } x \text{ from } 2\end{array}\right)\left(\begin{array}{c}\text{number of ways of} \\ \text{choosing } 2 - x \text{ from } 2\end{array}\right)}{\left(\begin{array}{c}\text{total number of ways of} \\ \text{choosing a sample of 2 from 4}\end{array}\right)}$$

$$= \frac{\dbinom{2}{x}\dbinom{2}{2-x}}{\dbinom{4}{2}} \qquad x = 0, \quad 1, \quad 2 \quad \text{i.e.,}$$

x	$p(x)$
0	$\dfrac{1}{6}$
1	$\dfrac{2}{3}$
2	$\dfrac{1}{6}$

b.

$$p(x) = \frac{\dbinom{1}{x}\dbinom{3}{2-x}}{\dbinom{4}{2}} \qquad x = 0, 1; \quad \text{i.e.,}$$

x	$p(x)$
0	$\dfrac{1}{2}$
1	$\dfrac{1}{2}$

c. $P(X = 0) = 1$

3.11   Let X = number on the ticket drawn and G_i = net gain for box i, i = I, II. Then
$G_i = X - 2$.

a. $E(G_I) = \sum\limits_{x=0}^{2} (x-1)p_I(x) = (-1)\left(\dfrac{1}{3}\right) + 0\left(\dfrac{1}{3}\right) + 1\left(\dfrac{1}{3}\right) = 0$

   $E(G_I^2) = \sum\limits_{x=0}^{2} (x-1)^2 p_I(x) = (1)\left(\dfrac{1}{3}\right) + 0\left(\dfrac{1}{3}\right) + 1\left(\dfrac{1}{3}\right) = \dfrac{2}{3}$

   $V(G_I) = E(G_I^2) - [E(G_I)]^2 = \dfrac{2}{3} - 0 = \dfrac{2}{3}$

b. $E(G_{II}) = \sum\limits_{x=0}^{2} (x-1)p_{II}(x) = (-1)\left(\dfrac{3}{5}\right) + 0\left(\dfrac{1}{5}\right) + 3\left(\dfrac{1}{5}\right) = 0$

   $E(G_{II}^2) = \sum\limits_{x=0}^{2} (x-1)^2 p_{II}(x) = (1)\left(\dfrac{3}{5}\right) + 0\left(\dfrac{1}{5}\right) + 9\left(\dfrac{1}{5}\right) = \dfrac{12}{5}$

   $V(G_{II}) = E(G_{II}^2) - [E(G_{II})]^2 = \dfrac{12}{5} - 0 = \dfrac{12}{5}$

c. Box II, since for Box I the highest possible net gain is \$1 with a probability of 1/3, but for Box II the highest possible gain is \$3 with a probability of 1/5. Note that $V(G_I) < V(G_{II})$; i.e., Box I net gain varies less from $E(G_i) = 0$ than Box II net gain.

3.13   Let X = age of death of a person infected with the AIDS virus from 1982−1989. Then we can estimate the mean and standard deviation of X from Figure 3.5 by letting the possible values of X be the mean ages for the different age groups in the pie chart. Hence,

$E(X) = 6P(X = 6) + 21P(X = 21) + 34.5P(X = 34.5)$

$+ 44.5P(X = 44.5) + 54.5P(X = 54.5) + 60P(X = 60)$

$= 6(0.014) + 21(0.194) + 34.5(0.45) + 44.5(0.221) + 54.5(0.082) + 60(0.038)$

$= 36.3$ (Answers may vary.)

$\text{Var}(X) = E(X^2) - (E(X))^2$

$E(X^2) = 36(0.014) + 441(0.194) + 1190.25(0.45) + 1980.25(0.221)$

$+ 2970(0.082) + 3600(0.038) = 1439.65$

$$\text{Var}(X) = 1439.65 - (36.3)^2 = 1439.65 - 1317.69 = 121.96 \text{ (Answers may vary.)}$$

$$\text{Std}(X) = (\text{Var}(X))^{(\frac{1}{2})} = (121.96)^{(\frac{1}{2})} = 11.04 \text{ (Answers may vary.)}$$

3.15 $\text{E(number of sales)} = 0 \cdot p(0) + 1 \cdot p(1) + 2 \cdot p(2) = 0(0.7) + 1(0.2) + 2(0.1) = 0.4$

$\text{V(number of sales)} = 0^2 \cdot p(0) + 1^2 \cdot p(1) + 2^2 \cdot p(2) - [\text{E(number of sales)}]^2$

$$= 0(0.7) + 1(0.2) + 4(0.1) - (0.4)^2 = 0.44$$

Standard deviation of Sales $= \sqrt{\text{V(sales)}} = \sqrt{0.44} = 0.6633.$

3.17

    a. Let X = weekly number of breakdowns. Using Tchebysheff's theorem, we have

$$P(\mu - k\sigma < X < \mu + k\sigma) \geq 1 - \tfrac{1}{k^2}$$

For $1 - \tfrac{1}{k^2} = 0.9$, we have $k = \sqrt{10}$, and thus the desired interval is

$$(\mu - k\sigma, \mu + k\sigma) = [4 - \sqrt{10}(0.8), \quad 4 + \sqrt{10}(0.8)] = (1.4702, \quad 6.5298).$$

    b. Eight breakdowns is $\dfrac{8 - \mu}{\sigma} = \dfrac{8 - 4}{0.8} = 5$ standard deviations from the mean.

The interval $(\mu - 5\sigma, \mu + 5\sigma)$ or $(0, 8)$ must contain at least $1 - \dfrac{1}{5^2} = 0.96$

of the probability. Thus, at most 4% of the probability mass can exceed 8 breakdowns and the director is safe in his claim.

3.19

    a. Let X = battery performance period. Using Tchebysheff's theorem, we have

$$P(\mu - k\sigma < X < \mu + k\sigma) \geq 1 - \tfrac{1}{k^2}$$

Solving $1 - \dfrac{1}{k^2} = 0.9$, for $k = \sqrt{10}$; thus the desired interval is

$$(\mu - \sqrt{10}\sigma, \mu + \sqrt{10}\sigma) = [100 - \sqrt{10}(5), \quad 100 + \sqrt{10}(5)] = (84.1886, \quad 115.8114).$$

b. No, since 80 $\not\subset$ (84.1886, 115.8114). Also, note that 80 is $\dfrac{80-100}{5} = -4$

standard deviations from the mean. Then $P(X \le \mu - 4\sigma) \le P(|X - \mu|$

$\ge 4\sigma) \le \dfrac{1}{4^2} = 0.0625$. Therefore, one would expect less than 6.25% of bat-

teries to die out in less than 80 minutes.

3.21

    a. $P(X = 2) = \dbinom{4}{2}(0.2)^2(0.8)^2 = 0.1536$

    b. $P(X \ge 2) = P(X = 2) + P(X = 3) + P(X = 4) = \sum\limits_{x=2}^{4} \dbinom{4}{x}(0.2)^x(0.8)^{4-x}$

        $= \dbinom{4}{2}(0.2)^2(0.8)^2 + \dbinom{4}{3}(0.2)^3(0.8) + \dbinom{4}{4}(0.2)^4$

        $= 0.1536 + 0.0256 + 0.0016 = 0.1808$

    c. $P(X \le 2) = P(X = 0) + P(X = 1) + P(X = 2) = \sum\limits_{x=0}^{2} \dbinom{4}{x}(0.2)^x(0.8)^{4-x}$

        $= \dbinom{4}{0}(0.8)^4 + \dbinom{4}{1}(0.2)^1(0.8)^3 + \dbinom{4}{2}(0.2)^2(0.8)^2$

        $= 0.4096 + 0.4096 + 0.1536 = 0.9728$

    d. $E(X) = np = 4(0.2) = 0.8$

    e. $V(X) = np(1 - p) = 4(0.2)(0.8) = 0.64$

3.23   Let X = number of underfilled boxes. Then X has a binomial distribution with parameters $n = 25$, p as given.

    a. $P(X \le 2) = 0.537$

    b. $P(X \le 2) = 0.098$

3.25

    a. $E(X) = np = 20(0.8) = 16$

    b. $V(X) = np(1 - p) = 20(0.8)(0.2) = 3.2$

3.27 $P(X \geq 5) = 1 - P(X \leq 4) > 0.9 \Rightarrow P(X \leq 4) < 0.1$

Using the formula

$$P(X \leq 4) = \sum_{x=0}^{4} \binom{n}{x} (0.8)^x (0.2)^{n-x}$$

we find

n	$P(X \leq 4)$
6	0.34464
7	0.14803
8	0.05628

Therefore, at least $n = 8$ people must donate blood for the probability of having at least 5 Rh+ donors to be greater than 0.9.

3.29   Let $X =$ number of firms out of a sample of five that say "quality of life" is an important factor. Then, assuming independence among firms, X has a binomial distribution with parameters $n = 5$, $p = 0.55$, and

$$P(X \geq 3) = \sum_{x=3}^{5} \binom{5}{x} (0.55)^x (0.45)^{5-x}$$

$$= \binom{5}{3} (0.55)^3 (0.45)^2 + \binom{5}{4} (0.55)^4 (0.45)^1 + \binom{5}{5} (0.55)^5$$

$$= 0.3369 + 0.2058 + 0.0503 = 0.5931.$$

3.31   Let $X =$ number of radar sets out of n that detect an intruding aircraft. Then X has binomial distribution with parameters n, and $p = 0.9$.

a. $P(X \geq 1) = 1 - P(X = 0) = 1 - \binom{2}{0} (0.9)^0 (0.1)^2 = 0.99$

b. $P(X \geq 1) = 1 - P(X = 0) = 1 - \binom{4}{0} (0.9)^0 (0.1)^4 = 0.9999$

3.33   Let $X =$ number of components out of the four that last longer than 1000 hours. The probability that a given component lasts longer than 1000 hours is 0.8; thus X has a binomial distribution with parameters $n = 4$, $p = 0.8$.

a. $P(X = 2) = \binom{4}{2}(0.8)^2(0.2)^2 = 0.1536$

b. $P(X \geq 2) = 1 - P(X \leq 1) = 1 - \binom{4}{0}(0.8)^0(0.2)^4 - \binom{4}{1}(0.8)^1(0.2)^3$

$= 1 - 0.0016 - 0.0256 = 0.9728$

3.35 Y has a binomial distribution with parameters $n = 4$, $p = 0.1$.

$E(C) = E(3Y^2 + Y + 2) = 3E(Y^2) + E(Y) + 2$

$= 3(V(Y) + [E(Y)]^2) + E(Y) + 2 = 3np(1 - p) + 3(np)^2 + np + 2$

$= 3(4)(0.1)(0.9) + 3[(4)(0.1)]^2 + 4(0.1) + 2 = 3.96$

3.37 Let $X =$ number of defective motors out of ten in the warehouse. Then X is binomially distributed with $n = 10$, $p = 0.08$. Let $Y =$ net gain $=$ (selling price for the ten motors) $-$ (twice the selling price of a motor) $\cdot X = 10(100) - 200X$

$= 1{,}000 - 200X$.

$E(Y) = E(1{,}000 - 200X) = 1{,}000 - 200E(X) = 1{,}000 - 200np = 1{,}000$

$-200(10)(0.08) = 840$

3.39

a. $P(Y \geq 4) = 1 - P(Y \leq 3) = 1 - P(Y = 2) - P(Y = 3)$

$= 1 - \binom{2-1}{2-1}(0.4)^2(0.6)^0 - \binom{3-1}{2-1}(0.4)^2(0.6)^1 = 1 - 0.16 - 2(0.4)^2(0.6)$

$= 0.648$

b. $P(Y = y)$ is nonzero only for $y = r$, $r + 1$, $r + 2$, $\ldots$. Therefore, for $r = 4$,

$P(Y \geq 4) = \sum_{y=4}^{\infty} p(y) = 1.$

3.41 Let $Y =$ the trial on which the third nondefective engine is found. Then Y has a negative binomial distribution, with $p = 0.9$, $r = 3$.

a. $P(Y = 5) = p(5) = \binom{y-1}{r-1}p^r(1-p)^{y-r} = \binom{4}{2}(0.9)^3(0.1)^2 = 0.04374$

b. $P(Y \le 5) = P(Y = 3) + P(Y = 4) + P(Y = 5) = p(3) + p(4) + p(5)$

$$= \binom{3-1}{3-1}(0.9)^3(0.1)^0 + \binom{4-1}{3-1}(0.9)^3(0.1)^1 + \binom{5-1}{3-1}(0.9)^3(0.1)^2$$

$$= 0.729 + 0.2187 + 0.04374 = 0.99144$$

3.43

a. Let Y be defined as in Exercise 3.40. Then Y has a geometric distribution with $p = 0.9$, and

$$E(Y) = \frac{1}{p} = \frac{10}{9}$$

$$V(Y) = \frac{1-p}{p^2} = \frac{0.1}{(0.9)^2} = 0.1235.$$

b. Let Y be defined as in Exercise 3.41. Then Y has a negative binomial distribution with parameters $p = 0.9$, $r = 3$, and

$$E(Y) = \frac{r}{p} = \frac{30}{9} = 3.33$$

$$V(Y) = \frac{r(1-p)}{p^2} = \frac{0.3}{(0.9)^2} = 0.3704.$$

3.45   Let total cost $= C$. Then $C = 20Y$, and

$$E(C) = 20E(Y) = 20\left(\frac{r}{p}\right) = 20\left(\frac{3}{0.4}\right) = 150.$$

$$V(C) = V(20Y) = 20^2V(Y) = 400\frac{r(1-p)}{p^2} = 400\frac{(3)(0.6)}{(0.4)^2} = 4{,}500$$

Standard deviation of $C = \sqrt{V(C)} = \sqrt{4{,}500} = 67.082$

Using Tchebysheff's theorem, we have $P(C > 350) = P(C - 150 > 350 - 150)$

$$\le P(|C - 150| \ge 200) = P\left(|C - 150| \ge \left(\frac{200}{67.082}\right)67.082\right)$$

$$\le \frac{1}{\left(\dfrac{200}{67.082}\right)^a} = 0.1125. \text{ Therefore, it is unlikely that the cost will exceed}$$

$350.

Note also that $P(C > 350)$ may be computed exactly as

$$P(C > 350) = P(20Y > 350) = P(Y > 17.5) = 1 - P(Y \le 17)$$

$$= 1 - \sum_{x=3}^{17} \binom{y-1}{3-1} (0.4)^3 (1 - 0.4)^{x-3}.$$

3.47

    a. Let $Y =$ number of the well in which oil was first struck. Then Y has a geometric distribution with parameter $p = 0.2$, so

$$P(Y = 3) = (1 - p)^{(3-1)} p^1 = (0.8)^2 (0.2)^1 = 0.128.$$

    b. Let $Y =$ number of the well in which the third oil strike occurs. Then Y has a negative binomial distribution with parameters $p = 0.2$, $r = 3$, so

$$P(Y = 5) = \binom{5-1}{3-1} (0.2)^3 (0.8)^2 = 0.03072.$$

The solutions to parts (a) and (b) require the assumption of independence of the wells.

3.49   Let $Y =$ number of tires that must be selected in order to get four good ones. Then Y has a negative binomial distribution with parameters $p = 0.9$, $r = 4$.

    a. $P(Y = 6) = \binom{6-1}{4-1} (0.9)^4 (0.1)^2 = 0.06561$

    b. $E(Y) = \dfrac{r}{p} = \dfrac{4}{0.9} = 4.4444$

    c. $V(Y) = \dfrac{r(1-p)}{p^2} = \dfrac{4(0.1)}{(0.9)^2} = 0.4938$

3.51

    a. Let $Y =$ number of customers it takes to sell the three white appliances. Then Y has a negative binomial distribution with parameters $p = \dfrac{1}{2}$, $r = 3$, and

$$P(Y = 5) = \binom{5-1}{3-1} \left(\frac{1}{2}\right)^3 \left(\frac{1}{2}\right)^{5-3} = 6 \left(\frac{1}{8}\right) \left(\frac{1}{4}\right) = \frac{3}{16}.$$

b. Let X = number of customers it takes to sell the brown appliances. Then X has the same distribution as Y, a negative binomial distribution with parameters $p = \frac{1}{2}$, $r = 3$, and $P(X = 5) = P(Y = 5) = \frac{3}{16}$.

c. $P(Y = 3) = \binom{3-1}{3-1} \left(\frac{1}{2}\right)^3 \left(\frac{1}{2}\right)^{3-3} = \left(\frac{1}{2}\right)^3 = \frac{1}{8}$

d. P(all the whites ordered before all browns) = $P(Y \leq 5)$

$$= p(3) + p(4) + p(5) = \frac{3}{16} + p(4) + \frac{1}{8}$$

$$= \frac{3}{16} + \binom{4-1}{3-1} \left(\frac{1}{2}\right)^3 \left(\frac{1}{2}\right) + \frac{1}{8}$$

$$= \frac{3}{16} + \frac{3}{16} + \frac{1}{8} = \frac{1}{2}$$

3.53 Let Y = number of calls arriving in a given one-minute period. Then Y has a Poisson distribution with parameter $\lambda = 4$.

a. $P(Y = 0) = p(0) = \frac{4^0}{0!} e^{-4} = e^{-4} = 0.0183$

b. $P(Y \geq 2) = 1 - P(Y \leq 1) = 1 - F(1) = 1 - 0.092 = 0.908$

c. Let X = number of calls arriving in a given two-minute period. Then X has a Poisson distribution with parameter $\lambda = 2(4) = 8$ and $P(X \geq 2)$
$= 1 - F(1) = 1 - 0.003 = 0.997$.

3.55 Let Y = number of fatalities per 10^9 vehicle miles with NMSL in effect. Then Y has a Poisson distribution with parameter $\lambda = 16$.

a. $P(Y \leq 15) = F(15) = 0.467$

b. $P(Y \geq 20) = 1 - P(Y \leq 19) = 1 - F(19) = 1 - 0.812 = 0.188$

3.57

    a. Let Y = number of teleport inquiries in one millisecond. Then Y has a Poisson distribution with parameter $\lambda = 0.2$ and

$$P(Y = 0) = \frac{(0.2)^0}{0!} e^{-0.2} = e^{-0.2} = 0.8187.$$

    b. Let X = number of teleport inquiries in three milliseconds. Then X has a Poisson distribution with parameter $\lambda = 3(0.2) = 0.6$ and

$$P(X = 0) = \frac{(0.6)^0}{0!} e^{-0.6} = e^{-0.6} = 0.5488.$$

3.59   Let Y = number of customer arrivals in a given hour. Then Y has a Poisson distribution with $\lambda = 8$.

    a. $P(Y = 8) = P(Y \leq 8) - P(Y \leq 7) = 0.593 - 0.453 = 0.140$

    b. $P(Y \leq 3) = 0.042$

    c. $P(Y \geq 2) = 1 - P(Y \leq 1) = 1 - 0.003 = 0.997$

3.61

    a. Let X = number of customers that arrive in a given two-hour period of time. Then X has a Poisson distribution with $\lambda = 2(8) = 16$ and

$$P(X = 2) = \frac{16^2}{2!} e^{-16} = 128 e^{-16} = 1.44 \times 10^{-5}.$$

    b. The two one-hour time periods are nonoverlapping, and therefore X = total number of customers that arrive in the given two-hour time period has a Poisson distribution with $\lambda = 2(8) = 16$, and, as for part (a), $P(X = 2)$
$= 1.44 \times 10^{-5}.$

Consistent with this answer, note the following. Let Y_1 = number of customers that arrive 1–2 pm, and Y_2 = number of customers that arrive 3–4 pm. Then Y_1 and Y_2 are each distributed as Poisson with $\lambda = 8$ and

$$P(Y_1 + Y_2 = 2) = P(Y_1 = 0, \ Y_2 = 2) + P(Y_1 = 1, \ Y_2 = 1) + P(Y_1 = 2, \ Y_2 = 0)$$

$$= 2 \cdot p(0)p(2) + [p(1)]^2 = 2\left(\frac{8^0}{0!}e^{-8}\right)\left(\frac{8^2}{2!}e^{-8}\right) + \left(\frac{8^1}{1!}e^{-8}\right)^2$$

$$= 7.2 \times 10^{-6} + 7.2 \times 10^{-6} = 1.44 \times 10^{-5}.$$

3.63 Let X = number of imperfections in an eight-square yard sample. Then X has a Poisson distribution with $\lambda = 8(4) = 32$. Let $C = 10X$ = cost of repair. Then

$$E(C) = 10E(X) = 10\lambda = 10(32) = 320$$

$$V(C) = (10)^2 V(X) = 100\lambda = 100(32) = 3{,}200$$

The standard deviation of $C = \sqrt{V(C)} = \sqrt{3{,}200} = 40\sqrt{2} = 56.5695$

3.65 $\displaystyle E[Y(Y-1)] = \sum_{y=0}^{\infty} y(y-1)\frac{\lambda^y e^{-\lambda}}{Y!} = \lambda^2 e^{-\lambda} \sum_{y=2}^{\infty} \frac{\lambda^{(y-2)}}{(y-2)!} = \lambda^2 e^{-\lambda} \sum_{x=0}^{\infty} \frac{\lambda^x}{x!}$

$$= \lambda^2 e^{-\lambda} e^{\lambda} = \lambda^2$$

$$V(Y) = E(Y^2) - [E(Y)]^2 = E(Y^2) - E(Y) + E(Y) - [E(Y)]^2$$

$$= E[Y(Y-1)] + E(Y) - [E(Y)]^2$$

$$= \lambda^2 + \lambda - (\lambda)^2 = \lambda$$

3.67

    a. Let Y = number of cars arriving in the first hour.

$$P(Y \geq 12) = 1 - P(Y \leq 11) = 1 - 0.999 = 0.001$$

b. Let X = the number of cars arriving in a given eight hours. Then X has a Poisson distribution with $\lambda = 8(4) = 32$ and

$$P(X \le 11) = \sum_{x=0}^{11} \frac{32^x e^{-32}}{x!} = e^{-32} \sum_{x=0}^{11} \frac{32^x}{x!}$$

$$= e^{-32} \left(1 + 32 + \frac{32^2}{2} + \frac{32^2}{6} + \cdots + \frac{32^{11}}{11!} \right)$$

$$= e^{-32}(1.345732 \times 10^9) = 0.000017.$$

3.69 Let Y = number of nondefectives in sample of 5. Then Y has a hypergeometric distribution with parameters $k = 6$, $n = 5$, N = 10, and

$$P(Y = 5) = p(5) = \frac{\binom{6}{5}\binom{4}{0}}{\binom{10}{5}} = \frac{1}{42}.$$

3.71 Let Y = number of local firms selected. Then Y has a hypergeometric distribution with parameters $k = 4$, $n = 3$, N = 6.

a. $P(\text{at least one not local}) = P(\text{not all local}) = 1 - P(Y = 3) = 1 - p(3)$

$$= 1 - \frac{\binom{4}{3}\binom{2}{0}}{\binom{6}{3}} = 1 - \frac{4}{20} = \frac{4}{5}$$

b.

$$P(Y = 3) = \frac{\binom{4}{3}\binom{2}{0}}{\binom{6}{3}} = \frac{1}{5}$$

3.73 Y has a hypergeometric distribution with parameters N = 10, $n = 3$, and k as given.

a. $k= 2$

y	$p(y)$
0	$\dfrac{\dbinom{2}{0}\dbinom{8}{3}}{\dbinom{10}{3}} = \dfrac{7}{15}$
1	$\dfrac{\dbinom{2}{1}\dbinom{8}{2}}{\dbinom{10}{3}} = \dfrac{7}{15}$
2	$\dfrac{\dbinom{2}{2}\dbinom{8}{1}}{\dbinom{10}{3}} = \dfrac{1}{15}$

b. $k= 4$

y	$p(y)$
0	$\dfrac{\dbinom{4}{0}\dbinom{6}{3}}{\dbinom{10}{3}} = \dfrac{1}{6}$
1	$\dfrac{\dbinom{4}{1}\dbinom{6}{2}}{\dbinom{10}{3}} = \dfrac{1}{2}$
2	$\dfrac{\dbinom{4}{2}\dbinom{6}{1}}{\dbinom{10}{3}} = \dfrac{3}{10}$
3	$\dfrac{\dbinom{4}{3}\dbinom{6}{0}}{\dbinom{10}{3}} = \dfrac{1}{30}$

3.75 Let Y = number of misfiring plugs among the four removed. Then Y has a hypergeometric distribution with $N = 8$, $n = 4$, $k = 2$, and

$$P(Y = 2) = \frac{\binom{2}{2}\binom{6}{2}}{\binom{8}{4}} = \frac{3}{14}.$$

3.77 Let Y = number of accounts past due that the auditor sees. Then Y has a hypergeometric distribution with $N = 8$, $n = 3$, k as given, and

$$P(Y \geq 1) = 1 - P(Y = 0) = 1 - \frac{\binom{k}{0}\binom{8-k}{3}}{\binom{8}{3}} = 1 - \frac{\binom{8-k}{3}}{56}.$$

a. $k = 2$; $P(Y \geq 1) = 1 - \dfrac{\binom{8-2}{3}}{56} = 1 - \dfrac{20}{56} = \dfrac{9}{14}$

b. $k = 4$; $P(Y \geq 1) = 1 - \dfrac{\binom{8-4}{3}}{56} = 1 - \dfrac{4}{56} = \dfrac{13}{14}$

c. $k = 7$; $P(Y \geq 1) = 1 - \dfrac{\binom{8-7}{3}}{56} = 1 - 0 = 1$, since the auditor must

choose at least two past-due accounts.

3.79 Note that Y has a hypergeometric distribution with parameters $N = 20$, $n = 5$, and k as given.

a. $k = 0$: $P(Y \leq 1) = 1$

b. $k = 1$: $P(Y \leq 1) = 1$

c. $k = 2$:

$$P(Y \leq 1) = p(0) + p(1) = \frac{\binom{2}{0}\binom{20-2}{5-0}}{\binom{20}{5}} + \frac{\binom{2}{1}\binom{20-2}{5-1}}{\binom{20}{5}} = \frac{21}{38} + \frac{15}{38} = \frac{18}{19}$$

d. $k = 3$:

$$P(Y \leq 1) = p(0) + p(1) = \frac{\binom{3}{0}\binom{17}{5}}{\binom{20}{5}} + \frac{\binom{3}{1}\binom{17}{4}}{\binom{20}{5}} = \frac{91}{228} + \frac{105}{228} = \frac{49}{57}$$

e. $k = 4$:

$$P(Y \leq 1) = p(0) + p(1) = \frac{\binom{4}{0}\binom{16}{5}}{\binom{20}{5}} + \frac{\binom{4}{1}\binom{16}{4}}{\binom{20}{5}} = \frac{1092}{3876} + \frac{1820}{3876} = \frac{728}{969}$$

3.81   Let Y = number of defectives from line I. Then Y has a hypergeometric distribution with parameters N = 10, n = 5, and

$$P(Y = 2) = \frac{\binom{4}{2}\binom{6}{3}}{\binom{10}{5}} = \frac{10}{21}.$$

3.83 $M(t) = E(e^{tY}) = \sum_{y=0}^{n} e^{ty} \binom{n}{y} p^y (1-p)^{(n-y)} = \sum_{y=0}^{n} \binom{n}{y} (pe^t)^y (1-p)^{(n-y)}$

$\quad\quad = [pe^t + (1-p)]^n$

since, using the binomial theorem, we have

$$\sum_{y=0}^{n} \binom{n}{y} a^y b^{(n-y)} = (a + b)^n$$

Therefore,

$$E(Y) = M'(0) = n[pe^t + (1 - p)]^{n-1} pe^t \big|_{t=0} = np$$

$$E(Y^2) = M''(0) = \left(n(n-1)[pe^t + (1-p)]^{n-1}(pe^t)^2 + n[pe^t + (1-p)]^{n-1}pe^t\right)\big|_{t=0}$$

$$= n(n-1)p^2 + np$$

$$V(Y) = E(Y^2) - [E(Y)]^2 = n(n-1)p^2 + np - (np)^2 = np(1 - p)$$

3.85 $M_Y(t) = E(e^{tY}) = E(e^{t(aX+b)}) = E[e^{tb}e^{(at)X}] = e^{tb}E[e^{(at)X}] = e^{tb}M_X(at)$

3.87   See Appendix for graph.

3.89   Let Y = number of radar sets out of the five that detect the plane. Then Y has a binomial distribution with parameters $n = 5$, $p = 0.9$.

$$P(Y = 4) = \binom{5}{4}(0.9)^4(0.1)^1 = 0.32805$$

and

$$P(Y \geq 1) = 1 - P(Y = 0) = 1 - \binom{5}{0}(0.9)^0(0.1)^5 = 0.99999$$

3.91 $P(Y \leq a) = \binom{5}{0}p^0(1 - p)^5 = (1 - p)^5$

     a. $(1 - p)^5 = (1)^5 = 1$

     b. $(1 - p)^5 = (0.9)^5 = 0.5905$

     c. $(1 - p)^5 = (0.7)^5 = 0.1681$

     d. $(1 - p)^5 = (0.5)^5 = 0.03125$

     e. $(1 - p)^5 = (0)^5 = 0$

3.93   For $n = 5$, $a = 1$, we have

p	$P(Y \le a)$
0.05	0.9774
0.10	0.9185
0.20	0.7373
0.30	0.5282
0.40	0.3370

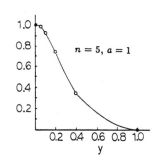

For $n = 25$, $a = 5$, we have

p	$P(Y \le a)$
0.05	0.9988
0.10	0.9666
0.20	0.6167
0.30	0.1935
0.40	0.0294

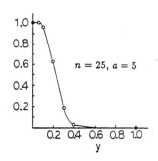

  a. $n = 25$, $a = 5$

  b. $n = 25$, $a = 5$

3.95   Let Y = number of colonies in a given dish. Then Y has a Poisson distribution with a mean of $\lambda = 12$

  a. $P(Y \ge 10) = 1 - P(Y \le 9) = 1 - 0.242 = 0.758$

  b. $E(Y) = \lambda = 12$, and the standard deviation of $Y = \sqrt{V(Y)} = \sqrt{\lambda} = \sqrt{12}$

     $= 3.4641$

  c. Using Tchebysheff's theorem, we have

$$P[E(Y) - 2\sqrt{V(Y)} < Y < E(Y) + 2\sqrt{V(Y)}] \ge 1 - \frac{1}{2^2} = 0.75$$

     I.e., the desired interval is $[12 - 2(3.4641), \quad 12 + 2(3.4641)] = (5.0718, \quad 18.9282)$.

3.97   Here, Y has a Poisson distribution with mean $\lambda p = 100(0.05) = 5$.

  a. $E(Y) = \lambda p = 5$

  b. $P(Y = 0) = 0.007$

  c. $P(Y > 5) = 1 - P(Y \le 5) = 1 - 0.616 = 0.384$

3.99 Let Y = number of left-turning vehicles out of n vehicles arriving while the light is red. Then Y has a binomial distribution with parameters $n = 5$, $p = 0.2$, $P(Y \le 3) = 0.993$. This number may be computed directly as

$$P(Y \le 3) = 1 - P(Y \ge 4) = 1 - P(Y = 4) - P(Y = 5)$$

$$= 1 - \binom{5}{4}(0.2)^4(0.8)^1 - \binom{5}{5}(0.2)^5(0.8)^0 = 0.993.$$

3.101

a. Using the binomial theorem, we have

$$\sum p(y) = \sum_{y=0}^{n} \binom{n}{y} p^y (1-p)^{n-y} = [p + (1-p)]^n = 1^n = 1$$

b. $\sum p(y) = \sum_{y=1}^{\infty} (1-p)^{y-1} p = p \sum_{y=1}^{\infty} (1-p)^{y-1} = p \sum_{y=0}^{\infty} (1-p)^y$

$$= p \left(\frac{1}{1 - (1-p)} \right) \qquad \left(\text{since } \sum_{x=0}^{\infty} a^x = \frac{1}{1-a} \quad \text{for} \quad a < 1 \right)$$

$$= p \left(\frac{1}{p} \right) = 1$$

c. Noting that $\sum_{x=0}^{\infty} \frac{\lambda^x}{x!} = e^\lambda$, we have $\sum p(y) = \sum_{y=0}^{\infty} \frac{\lambda^y e^{-\lambda}}{y!} = e^{-\lambda} \sum_{y=0}^{\infty} \frac{\lambda^y}{y!}$

$$= e^{-\lambda} e^\lambda = 1.$$

3.103 Let Y = total number of requests for welding units until the third brand-A unit is used. Then Y has a negative binomial distribution with parameters $r = 3$, $p = 0.7$, and

$$P(Y = 5) = \binom{5-1}{3-1}(0.7)^3(0.3)^{(5-3)} = 6(0.7)^3(0.3)^2 = 0.18522.$$

3.105 Let Y = number of people that have to be interviewed before encountering a consumer who prefers brand A. Then Y has a geometric distribution with parameter $p = 0.6$.

$$P(Y = 5) = (1-p)^{5-1} p = (0.4)^4(0.6) = 0.01536$$

and

$$P(Y \ge 5) = 1 - P(Y \le 4)$$

$$= 1 - p(1) - p(2) - p(3) - p(4)$$

$$= 1 - (0.6) - (0.4)(0.6) - (0.4)^2(0.6) - (0.4)^3(0.6)$$

$$= 1 - (0.6) - (0.24) - (0.096) - (0.0384)$$

$$= 0.0256$$

3.107 Note that Y has a binomial distribution with parameters $n = 1,000$, $p = 0.9$, so that

$$E(Y) = np = 1,000(0.9) = 900, \text{ and}$$

$$V(Y) = np(1 - p) = 1,000(0.9)(0.1) = 90.$$

Using Tchebysheff's theorem with $k = 2$, we have $P(\mu - 2\sigma < Y < \mu + 2\sigma) > 1 - \dfrac{1}{2^2}$; i.e.,

$$P(900 - 2\sqrt{90} < Y < 900 + 2\sqrt{90}) = P(881.026 < Y < 918.974) \geq 0.75.$$

3.109

a. Note that Y has a binomial distribution with parameters $n = 4$, $p = \dfrac{1}{3}$, and

distribution function $p(y) = \dbinom{4}{y} \left(\dfrac{1}{3}\right)^y \left(\dfrac{2}{3}\right)^{4-y}$.

y	$p(y)$
0	$\left(\dfrac{2}{3}\right)^4$
1	$4\left(\dfrac{1}{3}\right)\left(\dfrac{2}{3}\right)^3 = 2\left(\dfrac{2}{3}\right)^4$
2	$6\left(\dfrac{1}{3}\right)^2\left(\dfrac{2}{3}\right)^2 = \left(\dfrac{2}{3}\right)^3$
3	$4\left(\dfrac{1}{3}\right)^3\left(\dfrac{2}{3}\right)^1 = \dfrac{1}{3}\left(\dfrac{2}{3}\right)^3$
4	$\left(\dfrac{1}{3}\right)^4$

b. $P(Y \geq 3) = p(3) + p(4) = \frac{1}{3} \left(\frac{2}{3} \right)^3 + \left(\frac{1}{3} \right)^4 = \frac{1}{9}$

c. $E(Y) = np = 4 \left(\frac{1}{3} \right) = \frac{4}{3}$

d. $V(Y) = np(1 - p) = 4 \left(\frac{1}{3} \right) \left(\frac{2}{3} \right) = \frac{8}{9}$

3.111

a. Here, Y has a hypergeometric distribution with parameters $N = 100$, $n = 20$, $k = 40$, and

$$p(10) = \frac{\binom{40}{10} \binom{60}{10}}{\binom{100}{20}} = 0.1192.$$

b. Here, Y has a binomial distribution with parameters $n = 20$, $p = 0.40$.

$p(10) = F(10) - F(9) = 0.872 - 0.755 = 0.117$

Thus it appears that N is large enough that the binomial probability function is a good approximation to the hypergeometric probability function.

3.113 Let Y = number of items sold on a given day, P = daily profit, and X = number of items stocked. Note that for $Y \leq X$

$P = 1.2Y - X$

$E(P) = 1.2E(Y) - X$

and

$E(Y|X = 1) = 1$

$E(Y|X = 2) = 2$

$E(Y|X = 3) = 2p(2) + 3P(Y \geq 3) = 2(0.1) + 3(0.9) = 2.9$

$E(Y|X = 4) = 2p(2) + 3p(3) + 4p(4) = 2(0.1) + 3(0.4) + 4(0.5) = 3.4.$

Hence

$$E(P|X = 1) = 1.2(1) - 1 = 0.2$$
$$E(P|X = 2) = 1.2(2) - 2 = 0.4$$
$$E(P|X = 3) = 1.2(2.9) - 3 = 0.48$$
$$E(P|X = 4) = 1.2(3.4) - 4 = 0.08.$$

Therefore, expected profit is maximized at $X = 3$.

3.115   Number of combinations $= 26 \cdot 26 \cdot 10 \cdot 10 \cdot 10 \cdot 10 = 6{,}760{,}000$

$$E(\text{winnings}) = \$100{,}000 \left(\frac{1}{6{,}760{,}000} \right) + \$50{,}000 \left(\frac{2}{6{,}760{,}000} \right) + \$1{,}000 \left(\frac{10}{6{,}760{,}000} \right)$$

$$= \$0.031065$$

Therefore, it appears that the expected value of the coupon is considerably less than the price of a stamp. However, one might also consider what the probability of winning would be if the coupon weren't mailed back.

Chapter 4.

4.1

    a. Let Y = number of field plots out of ten that contain the insect. Y is a discrete random variable.

    b. Let Y = number of defects in a given sampled section. Let X = number of sampled sections containing at least 5 defects. Both Y and X are discrete random variables.

    c. Let Y = number of grains seen in a given cross section. Then Y is a discrete random variable.

    d. Let A = area proportion covered by grains of a certain size. A is a continuous random variable.

4.3

    a. $P(X > 3) = \int_3^\infty f(x)dx = \int_3^6 \frac{3}{32}(8x - x^2 - 12)dx = \frac{3}{32}\left(4x^2 - \frac{x^3}{3} - 12x\right)\Big|_3^6$

    $= \frac{27}{32} = 0.84375$

    b. $0.5 = P(X > b) = \int_b^6 \frac{3}{32}(8x - x^2 - 21)dx = \frac{3}{32}\left(4x^2 - \frac{x^3}{3} - 12x\right)\Big|_b^6$

    $= 0 - \frac{3}{32}\left(4b^2 - \frac{b^3}{3} - 12b\right)$

    i.e., b is the solution to $0 = b^3 - 12b^2 + 36b - 16$; i.e., $b = 4$. Alternatively, note that the density of X is symmetric about $x = 4$, so $P(X > 4) = P(X \leq 4) = \frac{1}{2}$.

4.5

    a.

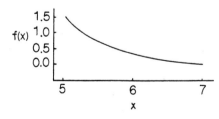

b. $F(x) = \begin{cases} 0 & x < 5 \\ \int_x^5 \frac{3}{8}(7-y)^2 dy = \frac{3}{8}\int_{-2}^{x-7} w^2 dw = \frac{w^3}{8}\Big|_{-2}^{x-7} = \frac{(x-7)^3}{8}+1 & 5 \le x \le 7 \\ 1 & x > 7 \end{cases}$

c. $P(X < 6) = F(6) = \frac{(6-7)^3}{8}+1 = \frac{7}{8}$

d. $P(X < 5.5|X < 6) = \frac{P(X < 5.5)}{P(X < 6)} = \frac{(5.5-7)^3 + 8}{8}\Big/\frac{7}{8} = \frac{37}{56}$

4.7

a. $P(X > \frac{1}{2}) = \int_{1/2}^1 2x\,dx = x^2\Big|_{1/2}^1 = 1 - \frac{1}{4} = \frac{3}{4}$

b. $P\left(X > \frac{1}{2}|X > \frac{1}{4}\right) = \frac{P\left(X > \frac{1}{2},\ X > \frac{1}{4}\right)}{P\left(X > \frac{1}{4}\right)} = \frac{P\left(> \frac{1}{2}\right)}{P\left(X > \frac{1}{4}\right)}$

$= \frac{\frac{3}{4}}{\int_{1/4}^1 2x\,dx} = \frac{4}{5}$

c. $P\left(X > \frac{1}{4}|X > \frac{1}{2}\right) = \frac{P\left(X > \frac{1}{4},\ X > \frac{1}{2}\right)}{P\left(X > \frac{1}{2}\right)} = \frac{P\left(X > \frac{1}{2}\right)}{P\left(X > \frac{1}{2}\right)} = 1$

d. $F(x) = \begin{cases} 0 & x < 0 \\ \int_0^x 2y\,dy = x^2 & 0 \le x \le 1 \\ 1 & x > 1 \end{cases}$

Yes, $F(x)$ is continuous.

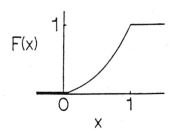

4.9 $E(X) = \int_{59}^{61} x f(x) dx = \int_{59}^{61} \frac{x}{2} dx = \frac{x^2}{4}\Big|_{59}^{61} = \frac{1}{4}[(61)^2 - (59)^2] = 60$

$$E(X^2) = \int_{59}^{61} x^2 f(x) dx = \int_{59}^{61} \frac{x^2}{2} dx = \frac{x^3}{6}\Big|_{59}^{61} = \frac{1}{6}[(61)^3 - (59)^3] = \frac{21,602}{6}$$

$$V(X) = E(X^2) - [E(X)]^2 = \frac{21,602}{6} - (60)^2 = \frac{1}{3}$$

4.11 $E(X) = \int_{-\infty}^{\infty} x f(x) dx = \frac{3}{32} \int_{2}^{6} x(x-2)(6-x) dx = -\frac{3}{32}\left(\frac{x^4}{4} - \frac{8x^3}{3} + 6x^2\right)\Big|_{2}^{6}$

$$= 4 \text{ hundred calories}$$

4.13

   a. $E(X) = \frac{3}{8}\int_{5}^{7} x(7-x)^2 dx = \frac{3}{8}\left(\frac{49x^2}{2} - \frac{14x^3}{3} + \frac{x^4}{4}\right)\Big|_{5}^{7} = 5.5$

     $E(X^2) = \frac{3}{8}\int_{5}^{7} x^2(7-x)^2 dx = \frac{3}{8}\left(\frac{49x^3}{3} - \frac{7x^4}{2} + \frac{x^5}{5}\right)\Big|_{5}^{7} = 30.4$

     $V(X) = E(X^2) - [E(X)]^2 = 30.4 - (5.5)^2 = 0.15$

   b. Using Tchebysheff's theorem, with k = 2, we have

$$P[E(X) - 2\sqrt{V(X)} < X < E(X) + 2\sqrt{V(X)}] \geq 1 - \frac{1}{(2)^2} = 0.75$$

     so the desired interval is $E(X) \pm 2\sqrt{V(X)} = (5.5 \pm 2\sqrt{0.15}) = (4.7254, 6.2746)$.

   c. $P(X < 5.5) = \int_{-\infty}^{5.5} f(x) dx = \int_{5}^{5.5} \frac{3}{8}(7-x)^2 dx = \frac{3}{8}\left(49x - 7x^2 + \frac{x^3}{3}\right)\Big|_{5}^{5.5}$

     $= 0.5781$

     We would expect to see about 58% of the pH measurements to be below 5.5.

4.15

   a. $F(x) = \begin{cases} 0 & x < a \\ \int_{a}^{x} \frac{1}{b-a} dy = \frac{x-a}{b-a} & a \leq x \leq b \\ 1 & x > b \end{cases}$

b. $P(X > c) = \int_c^b \frac{1}{b-a} dy = \frac{b-c}{b-a}$

c. $P(X > d | X > c) = \frac{P(X > d, X > c)}{P(X > c)} = \frac{P(X > d)}{P(X > c)} = \frac{\dfrac{b-d}{b-a}}{\dfrac{b-c}{b-a}} = \frac{b-d}{b-c}$

4.17 Here, X has a uniform distribution with parameters a = 0, b = 500.

a. $P(X > 475) = \int_{475}^{500} \frac{1}{500-0} dx = \frac{500-475}{500} = \frac{1}{20}$

b. $P(X < 25) = \int_0^{25} \frac{1}{500-0} dx = \frac{20-0}{500} = \frac{1}{20}$

c. $P(X < 250) = \int_0^{250} \frac{1}{500-0} dx = \frac{250-0}{500} = \frac{1}{2}$

4.19 Let X = time in seconds that the call arrived. Then X has a uniform distribution with a = 0, b = 60, and $P(X > 15) = \int_{15}^{60} \frac{1}{60-0} dx = \frac{60-15}{60} = \frac{3}{4}$.

4.21 Let X = hour of operation in which the defective board was produced. Since the number of defectives is Poisson, then, given that one defective was produced, the actual time of occurrence is equally likely in any small subinterval of time of a given size, and thus X has a uniform distribution with parameters a = 0, b = 8.

a. $P(X < 1) = \int_0^1 \frac{1}{8} dx = \frac{1}{8}$

b. $P(X > 7) = \int_7^8 \frac{1}{8} dx = \frac{1}{8}$

c. $P(4 < X \le 5 | X > 4) = \frac{P(4 < X \le 5, X > 4)}{P(X > 4)} = \frac{P(4 < X \le 5)}{P(X > 4)} = \frac{\int_4^5 \frac{1}{8} dx}{\int_4^8 \frac{1}{8} dx}$

$= \frac{1}{4}$

4.23 Let X = measurement error. Then X has a uniform distribution with parameters
$a = -0.02, b = 0.05$.

a. $P(-0.01 < X < 0.01) = \int_{-0.01}^{0.01} \dfrac{1}{0.05 - (-0.02)} dx = \dfrac{2}{7}$

b. $E(X) = \dfrac{a+b}{2} = \dfrac{0.05 + (-0.02)}{2} = 0.015$

$V(X) = \dfrac{(b-a)^2}{12} = \dfrac{[0.05 - (-0.02)]^2}{12} = \dfrac{49}{120,000} = 0.0004083.$

4.25 Let X = time of arrival measured from the beginning of the 30-minute period.
Since the number of arrivals is Poisson, the time of the arrival is equally likely in
any subinterval of time of a given size in the 30 minutes, and thus X has a uniform
distribution with parameters a = 0, b = 30, and

$P(X > 25) = \int_{25}^{30} \dfrac{1}{30} dx = \dfrac{30 - 25}{30} = \dfrac{1}{6}.$

4.27 Let X = stopping distance. Then X has a uniform distribution with parameters
a and b.

a. $P(X - a < b - X) = P\left(X < \dfrac{a+b}{2}\right) = \int_a^{a+b/2} \dfrac{1}{b-a} dx = \dfrac{\dfrac{a+b}{2} - a}{b-a} = \dfrac{1}{2}$

b. $P[X - a > 3(b - X)] = P\left(X > \dfrac{3b+a}{4}\right) = \int_{(3b+a)/4}^{b} \dfrac{1}{b-a} dx = \dfrac{b - \dfrac{3b+a}{4}}{b-a}$

$= \dfrac{1}{4}$

4.29 Let X = cycle time. Then X has a uniform distribution with parameters $a = 50$,
$b = 70$.

a. $E(X) = \dfrac{a+b}{2} = \dfrac{50 + 70}{2} = 60$

$V(X) = \dfrac{(b-a)^2}{12} = \dfrac{(70 - 50)^2}{12} = \dfrac{100}{3}$

b. Let T = number of trucks needed. Then we have $E(X)/T = 15$; i.e., $60/15 = 4$ trucks are needed.

4.31 Let X = magnitude of the next earthquake.

a. $P(X > 3) = 1 - F(3) = 1 - (1 - e^{-3/\theta}) = e^{-3/2.4} = e^{-5/4} = 0.2865$.

b. $P(2 < X < 3) = F(3) - F(2) = 1 - e^{-3/2.4} - (1 - e^{-2/2.4}) = e^{-5/6} - e^{-5/4}$
$= 0.1481$.

4.33 Let X = water demand. Then X has an exponential distribution with parameter $\theta = 100$.

a. $P(X > 200) = 1 - F(200) = 1 - (1 - e^{-200/100}) = e^{-2} = 0.1353$

b. Let k = maximum water-producing capacity. Then $0.01 = P(X > k)$

$= 1 - F(k)$

$= 1 - (1 - e^{-k/100}) = e^{-k/100}$. Hence $k = -100 \ln 0.01$

$= 460.52$ cfs.

4.35

a. Note that since $\int_0^\infty x^{(n-1)} e^{-(x/\theta)} dx = \Gamma(n)\theta^n$, then for k integer valued we have

$$E(X^k) = \frac{1}{\theta} \int_0^\infty x^k e^{-x/\theta} dx = \frac{1}{\theta}\Gamma(k+1)\theta^{(k+1)} = \theta^k k!$$

So

$E(X) = (10)1! = 10$

$E(X^2) = (10)^2 2! = 200$

$E(X^3) = (10)^3 3! = 6,000$

$E(X^4) = (10)^4 4! = 240,000$

and

$$E(C) = 100 + 40E(X) + 3E(X^2) = 100 + 40(10) + 3(200) = 1{,}100$$

$$E(C^2) = E\left\{[100 + 40(X) + 3(X^2)]^2\right\}$$

$$= 10{,}000 + 8{,}000\ E(X) + 2{,}200\ E(X^2) + 240E(X^3) + 9E(X^4)$$

$$= 10{,}000 + 8{,}000(10) + 2{,}200(200) + 240(6{,}000) + 9(240{,}000)$$

$$= 4{,}130{,}000$$

$$V(C) = E(C^2) - [E(C)]^2 = 4{,}130{,}000 - (1{,}100)^2 = 2{,}920{,}000$$

b. $P(C > 2{,}000) = P(3X^2 + 40X + 100 > 2{,}000) = P(3X^2 + 40X - 1{,}900 > 0)$

$$= P[(X - r_1)(X - r_2) > 0]$$

where $r_1 = \dfrac{10}{3}(-2 + \sqrt{61}) = 19.3675$, and $r_2 = \dfrac{10}{3}(-2 - \sqrt{61}) = -32.7$.

Therefore

$$P(C > 2{,}000) = P(X - r_1 > 0, X - r_2 > 0) + P(X - r_1 < 0, X - r_2 < 0)$$

$$= P(X > r_1, X > r_2) + P(X < r_1, X < r_2) = P(X > r_1) + P(X < r_2)$$

$$= PX(> r_1) = 1 - (1 - e^{-r_1/10}) = e^{-1.93675} = 0.1442$$

4.37 Let $X = $ tire lifelength. Then X has an exponential distribution with parameter $\theta = 30$.

  a. $P(X > 30) = 1 - F(30) = 1 - (1 - e^{-30/30}) = e^{-1} = 0.3679$

  b. Using the result of Exercise 4.30, we have

$$P(X > 30 | X > 15) = P(X > 30 - 15)$$

$$= 1 - F(15) = 1 - (1 - e^{-15/30}) = e^{-1/2} = 0.6065.$$

4.39 If the number of breakdowns has a Poisson distribution with parameter $\lambda = 0.5$, then the waiting time between breakdowns, X, has an exponential distribution with parameter $\theta = \dfrac{1}{\lambda} = \dfrac{1}{0.5} = 2$.

a. $P(X > 1) = 1 - F(1) = 1 - (1 - e^{-1/2}) = e^{-1/2} = 0.6065$

b. $P\left(X > \dfrac{1}{2}\right) = 1 - (1 - e^{-0.5/2}) = e^{-1/4} = 0.7788$

c. No, because of the result of Exercise 4.30, the "memoryless" property of the exponential distribution.

4.41   Let X = the weekly rainfall totals.

a. $P(X > 2) = 1 - F(2) = 1 - (1 - e^{-2/1.6}) = e^{-5/4} = 0.2865$

b. Let Y = number of weeks out of the next two in which rainfall doesn't exceed 2 inches. Then Y has a binomial distribution with parameters n = 2, p = $P(X \le 2) = F(2) = 1 - e^{-5/4}$. Then

$$P(Y = 2) = \binom{2}{2}\left(1 - e^{-5/4}\right)^2 \left(e^{-5/4}\right)^0 (1 - e^{-5/4})^2 = 0.5091.$$

4.43   Let X = repair time.

a. $P(X > 10) = F(10) = 1 - e^{-10/22} = 0.3653$

b. $P(30 < X < 60) = F(60) - F(30) = 1 - e^{-60/22} - (1 - e^{-30/22}) = e^{-15/11}$

$-e^{-30/11} = 0.1903$

c. Let k = the desired constant such that $P(X > k) = 1 - F(k) = 1 - (1$

$-e^{-k/22}) = e^{-k/22} = 0.10$. Solving for k, we have $k = -22 \ln (0.10) = 50.66$ minutes.

4.45

a. Let X = summer rainfall total.

$E(X) = \alpha\beta = 1.6(2) = 3.2$

$V(X) = \alpha\beta^2 = 1.6(2)^2 = 6.4$

b. Using Tchebysheff's theorem with k = 2, we have 0.75

$$= 1 - \frac{1}{2^2} \le P\left[E(X) - 2\sqrt{V(X)} < X < E(X) + 2\sqrt{V(X)}\right]$$

$$= P(3.2 - 2\sqrt{6.4} < X < 3.2 + 2\sqrt{6.4})$$

$$= P(-1.86 < X < 8.26). \text{ Since rainfall totals are nonnegative, we have the}$$

interval (0, 8.26).

4.47 For k integer valued, note that

$$E(Y^k) = \int_0^\infty \frac{1}{\Gamma(\alpha)\beta^\alpha} y^{(\alpha+k-1)} e^{-y/\beta} dy$$

$$= \frac{\beta^k \Gamma(\alpha+k)}{\Gamma(\alpha)} \int_0^\infty \frac{1}{\Gamma(\alpha+k)\beta^{\alpha+k}} y^{(\alpha+k-1)} e^{-y/\beta} dy = \frac{\beta^k \Gamma(\alpha+k)}{\Gamma(\alpha)}$$

a. $E(L) = 30E(Y) + 2E(Y^2) = 30\alpha\beta + \dfrac{2\beta^2\Gamma(\alpha+2)}{\Gamma(\alpha)} = 30\alpha\beta + \dfrac{2\beta^2(\alpha+1)!}{(\alpha-1)!}$

$$= 30(3)(2) + \frac{2(2)^2(24)}{(2)} = 276$$

$$V(L) = E(L^2) - [E(L)]^2 = E[(30Y + 2Y^2)^2] - (276)^2$$

$$= 900E(Y^2) + 120E(Y^3) + 4E(Y^4) - (276)^2$$

$$= 900(2)^2\frac{\Gamma(5)}{\Gamma(3)} + 120(2)^3\frac{\Gamma(6)}{\Gamma(3)} + 4(2)^4\frac{\Gamma(7)}{\Gamma(3)} - (276)^2$$

$$= 900(4)\frac{24}{2} + 120(8)\frac{120}{2} + 4(16)\frac{720}{2} - (276)^2 = 47,664$$

b. Using Tchebysheff's theorem, we want k such that $1 - \dfrac{1}{k^2} = 0.89$; i.e.,

$k \approx 3$. Then the desired interval is $\left[E(L) - 3\sqrt{V(L)}, \quad E(L) + 3\sqrt{V(L)}\right] =$

$(276 - 3\sqrt{47,664}, \quad 276 + 3\sqrt{47,664}) = (-378.963, \quad 930.963)$. Since L is

nonnegative, the interval is $(0, 930.963)$.

4.49 Let X_i = time to completion of the given task, $i = 1, 2$. Then X_i has a Gamma

distribution with parameters $\alpha = 1, \quad \beta = 10$, and $Y = X_1 + X_2$ has a Gamma

distribution with parameters $\alpha = 2(1) = 2, \quad \beta = 10$.

a. $E(Y) = \alpha\beta = 2(10) = 20$ and $V(Y) = \alpha\beta^2 = 2(10)^2 = 200$

b. Let A = average time to completion of the two tasks $= \dfrac{X_1 + X_2}{2} = \dfrac{Y}{2}$. Then

$$E(A) = \frac{1}{2}E(Y) = \frac{1}{2}(20) = 10 \text{ and } V(A) = \left(\frac{1}{2}\right)^2 V(Y) = \frac{1}{4}(200) = 50$$

4.51

    a. Let Y = maximum river flow. Then Y has a Gamma distribution with parameters $\alpha = 1.6$, $\beta = 150$. Therefore we have

$$E(Y) = \alpha\beta = 1.6(150) = 240$$

$$V(Y) = \alpha\beta^2 = 1.6(150)^2 = 36,000$$

Std dev $(Y) = \sqrt{V(Y)} = \sqrt{36,000} = 189.74$.

    b. Using Tchebysheff's theorem, we want k such that $1 - \dfrac{1}{k^2} = \dfrac{8}{9}$; i.e., $k = 3$.

Then the desired interval is $[E(Y) - 3\sqrt{V(Y)}, \quad E(Y) + 3\sqrt{V(Y)}] = (240 - 3\sqrt{36,000}, \quad 240 + 3\sqrt{36,000}) = (-329.21, \quad 809.21)$. Since Y is nonnegative, the interval is $(0, 809.21)$.

4.53   Since service times have a Gamma distribution with parameters $\alpha = 1$, $\beta = 3.2$, then the service time for three waiting customers, Y, has a Gamma distribution with parameters $\alpha = 3(1) = 3, \beta = 3.2$, and

$$E(Y) = \alpha\beta = 3(3.2) = 9.6$$

$$V(Y) = \alpha\beta^2 = 3(3.2)^2 = 30.72$$

$$f(y) = \begin{cases} \dfrac{1}{3.2^3\Gamma(3)}y^2 e^{-x/3.2} = \dfrac{1}{65.536}y^2 e^{-y/3.2} & y > 0 \\ 0 & y \leq 0. \end{cases}$$

4.55

    a. $P(0 \leq Z \leq 1.2) = 0.3849$

    b. $P(-0.9 \leq Z \leq 0) = P(0 \leq Z \leq 0.9) = 0.3159$

    c. $P(0.3 \leq Z \leq 1.56) = P(0 \leq Z \leq 1.56) - P(0 \leq Z \leq 0.3) = 0.4406$
       $- \quad 0.1179 = 0.3227$

    d. $P(-0.2 \leq Z \leq 0.2) = 2P(0 \leq Z \leq 0.2) = 2(0.0793) = 0.1586$

    e. $P(-2.00 \leq Z \leq -1.56) = P(1.56 \leq Z \leq 2.00) = P(0 \leq Z \leq 2.00)$
       $- \quad P(0 \leq Z \leq 1.56)$
       $= 0.4772 - 0.4406 = 0.0366$

4.57 Let X = amount spent on maintenance and repairs. Then X has a normal distribution with parameters $\mu = 400, \sigma = 20$, and

$$P(X > 450) = P\left(\frac{X - 400}{20} > \frac{450 - 400}{20}\right) = P(Z > 2.5) = 0.5 - 0.4938$$

$$= 0.0062.$$

4.59 Let X = diameter. Then X has a normal distribution with parameters $\mu = 1.005, \sigma = 0.01$, and $P(X < 0.98) + P(X > 1.02) = P\left(Z < \frac{0.98 - 1.005}{0.01}\right)$

$$+ \quad P\left(Z > \frac{1.02 - 1.005}{0.01}\right)$$

$$= P(Z < -2.5) + P(Z > 1.5) = (0.5 - 0.4938) + (0.5 - 0.4332)$$

$$= 0.0730.$$

4.61 Let X = resistances of wires produced by Company A. Then X has a normal distribution with parameters $\mu = 0.13, \quad \sigma = 0.005$.

a. $P(0.12 < X < 0.14) = P\left(\dfrac{0.12 - 0.13}{0.005} < \dfrac{X - 0.13}{0.005} < \dfrac{0.14 - 0.13}{0.005}\right)$

$$= P(-2 < Z < 2) = 2P(0 < Z < 2) = 2(0.4772) = 0.9544$$

b. Let Y = number of wires of a sample of four from Company A that meet specifications. Then Y has a binomial distribution with parameters $n = 4$, $p = 0.9544$, and

$$P(Y = 4) = \binom{4}{4} (0.9544)^4 (1 - 0.9544)^0 = 0.8297.$$

4.63 $P(|X| > 5) = P(X < -5) + P(X > 5) = P\left(Z < \dfrac{-5 - 0}{10}\right) + P\left(Z > \dfrac{5 - 0}{10}\right)$

$$= 2P(Z > 0.5) = 2(0.5 - 0.1915) = 0.6170$$

$$P(|X| > 10) = P(X < -10) + P(X > 10) = P\left(Z < \dfrac{-10 - 0}{10}\right) + P\left(Z > \dfrac{10 - 0}{10}\right)$$

$$= 2P(Z > 1) = 2(0.5 - 0.3413) = 0.3174$$

4.65   Let $X =$ monthly sickleave time. Then X has a normal distribution with parameters $\mu = 200,\quad \sigma = 20$.

a. $P(X < 150) = P\left(\dfrac{X - 200}{20} < \dfrac{150 - 200}{20}\right) = P(Z < -2.5) = 0.5 - 0.4938$

$= 0.0062$

b. Let $x_0 =$ desired time budgeted. Then $P(X > x_0) = P(Z > \dfrac{x_0 - 200}{20}) \overset{\text{set}}{=} 0.1$.

Note that $P(X > 1.28) = 0.1$, so we have $\dfrac{x_0 - 200}{20} = 1.28$, i.e., $x_0 = 225.6$ hours.

4.67   Let $X =$ amount of fill per box. Then X has a normal distribution with parameters

$\mu, \sigma = 1$, and $P(X > 16) = P\left(Z\dfrac{16 - \mu}{1}\right) \overset{\text{set}}{=} 0.01$. Note that $P(Z > 2.33) = 0.01$,

so we have $\dfrac{16 - \mu}{1} = 2.33$; i.e., $\mu = 13.67$ ounces.

4.69

a. Yes, it does appear that the total points can be modeled by a normal distribution.

b. According to the empirical rule, 68% of the data should lie one standard deviation above and below the mean and 95% of the data should lie within two standard deviations above and below the mean. Hence, consider the interval $(\bar{x} - s,\quad \bar{x} + s) = (143 - 26,\quad 143 + 26) = (117, 169)$. Notice that more than 77% of the games had total scores within $(117, 169)$. Now consider the interval $(\bar{x} - 2s,\quad \bar{x} + 2s) = (143 - 2(26),\quad 143 + 2(26)) = (91,\quad 195)$. Notice that less than 5% of the total scores fall outside of this region.

c. No and no. A score of 200 is greater than two standard deviations away from the mean. Such a score should occur less than 2.5% of the time, according to the empirical rule. A score of 250 is greater than three standard deviations away from the mean, making it even less likely to occur.

d. About 4 games.

4.71

    a. Discussion question. (Q–Q plot is in the Appendix.)

    b. $\bar{x} \approx 64 \qquad s \approx 9.5$

4.73

    a. $1 = \int_0^1 kx^3(1-x)^2 dx = k\int_0^1 x^{(4-1)}(1-x)^{(3-1)}dx = k\dfrac{\Gamma(4)\Gamma(3)}{\Gamma(7)} = k\dfrac{6(2)}{720}$

    $= k\dfrac{1}{60}$; i.e., $k = \dfrac{\Gamma(7)}{\Gamma(4)\Gamma(3)} = 60$, and X has a Beta distribution with parame-

    ters $\alpha = 4, \quad \beta = 3$.

    b. $E(X) = \dfrac{\alpha}{\alpha + \beta} = \dfrac{4}{7}$

    $V(X) = \dfrac{\alpha\beta}{(\alpha+\beta)^2(\alpha+\beta+1)} = \dfrac{4(3)}{(7)^2(8)} = \dfrac{3}{98}$

4.75   Note that

$$E(X^k) = \int_0^1 x^k \frac{\Gamma(\alpha+\beta)}{\Gamma(\alpha)\Gamma(\beta)}x^{\alpha-1}(1-x)^{\beta-1}dx = \int_0^1 \frac{\Gamma(\alpha+\beta)}{\Gamma(\alpha)\Gamma(\beta)}x^{\alpha+k-1}(1-x)^{\beta-1}dx$$

$$= \frac{\Gamma(\alpha+\beta)\Gamma(\alpha+k)}{\Gamma(\alpha)\Gamma(\alpha+k+\beta)}\int_0^1 \frac{\Gamma(\alpha+\beta+k)}{\Gamma(\alpha+k)\Gamma(\beta)}x^{\alpha+k-1}(1-x)^{\beta-1}dx$$

$$= \frac{\Gamma(\alpha+\beta)\Gamma(\alpha+k)}{\Gamma(\alpha)\Gamma(\alpha+k+\beta)}.$$

    a. $E(C) = 10 + 20E(X) + 4E(X^2) = 10 + (20)\dfrac{\alpha}{\alpha+\beta} + (4)\dfrac{\Gamma(\alpha+\beta)\Gamma(\alpha+2)}{\Gamma(\alpha)\Gamma(\alpha+2+\beta)}$

    $= 10 + (20)\dfrac{1}{3} + (4)\dfrac{\Gamma(3)\Gamma(3)}{\Gamma(1)\Gamma(5)} = 10 + \dfrac{20}{3} + (4)\dfrac{(2)(2)}{(24)} = \dfrac{52}{3} = 17.33$

    $E(C^2) = E[10 + 20X + 4(X^2)]^2 = 100 + 400E(X) + 480E(X^2) + 160E(X^3)$

    $+ 16E(X^4)$

    $= 100 + (400)\dfrac{1}{3} + (480)\dfrac{1}{6} + 160\dfrac{\Gamma(3)\Gamma(4)}{\Gamma(1)\Gamma(6)} + (16)\dfrac{\Gamma(3)\Gamma(5)}{\Gamma(1)\Gamma(7)}$

    $= 100 + \dfrac{400}{3} + \dfrac{480}{6} + \dfrac{160}{10} + \dfrac{16}{15} = \dfrac{1,652}{5}$

Then we have

$$V(C) = E(C^2) - [E(C)]^2 = \frac{1,652}{5} - \left(\frac{52}{3}\right)^2 = \frac{1,348}{45} = 29.96.$$

b. Using Tchebysheff's theorem with $k = 2$, we have $1 - \frac{1}{k^2} = \frac{3}{4}$, so the desired interval is

$$[E(C) - 2\sqrt{V(C)}, \quad E(C) + 2\sqrt{V(C)}] = \left(\frac{52}{3} - 2\sqrt{\frac{1348}{45}}, \quad \frac{52}{3} + 2\sqrt{\frac{1348}{45}}\right)$$

$$= (6.387, \quad 28.280).$$

4.77 $E(X) = \dfrac{\alpha}{\alpha + \beta} = \dfrac{4}{4 + 2} = \dfrac{2}{3}$. The angle corresponding to E(X) is $360° E(X)$

$$= 360° \left(\frac{2}{3}\right) = 240°.$$

4.79

a. $E(X) = \dfrac{\alpha}{\alpha + \beta} = \dfrac{3}{6} = \dfrac{1}{2}$

$V(X) = \dfrac{\alpha\beta}{(\alpha + \beta)^2(\alpha + \beta + 1)} = \dfrac{9}{(3 + 3)^2(3 + 3 + 1)} = \dfrac{1}{28}$

b. $E(X) = \dfrac{\alpha}{\alpha + \beta} = \dfrac{2}{4} = \dfrac{1}{2}$

$V(X) = \dfrac{\alpha\beta}{(\alpha + \beta)^2(\alpha + \beta + 1)} = \dfrac{4}{(2 + 2)^2(2 + 2 + 1)} = \dfrac{1}{20}$

c. $E(X) = \dfrac{\alpha}{\alpha + \beta} = \dfrac{1}{2}$

$V(X) = \dfrac{\alpha\beta}{(\alpha + \beta)^2(\alpha + \beta + 1)} = \dfrac{1}{(1 + 1)^2(1 + 1 + 1)} = \dfrac{1}{12}$

d. Case (a) exemplifies the best blending since it has the smallest variance.

4.81   Let X = fatigue life. Then X has a Weibull distribution with parameters $\gamma = 2$, $\theta = 4$.

    a. $P(X < 2) = F(2) = 1 - e^{-2^2/4} = 1 - e^{-1} = 0.6321$

    b. $E(X) = \theta^{1/\gamma}\Gamma\left(1 + \dfrac{1}{\gamma}\right) = 4^{1/2}\Gamma\left(1 + \dfrac{1}{2}\right) = 2\left(\dfrac{1}{2}\right)\Gamma\left(\dfrac{1}{2}\right) = \Gamma\left(\dfrac{1}{2}\right) = \sqrt{\pi}$

4.83   Let X = the time necessary to achieve proper blending of copper powders. Then X has a Weibull distribution with parameters $\gamma = 1.1$, $\theta = 2$, and

$$P(X < 2) = F(2) = 1 - e^{-(2)^{1.1}/2} = 0.6576.$$

4.85   Let X = steel-beam yield strength. Then X has a Weibull distribution with parameters $\gamma = 2$, $\theta = 3,600$. Let Y = number of beams out of two that have yield strengths greater than 70,000 psi. Then Y has a binomial distribution with parameters $n = 2$, $p = P(X > 70) = 1 - F(70) = 1 - \left(1 - e^{-(70)^2/3600}\right) = e^{-49/36}$, and

$$P(Y = 2) = \binom{2}{2}\left(e^{-49/36}\right)^2\left(1 - e^{-49/36}\right)^0 = 0.06573.$$

4.87   Let X = resistor lifelength in thousands of hours. Then X has a Weibull distribution with parameters $\gamma = 2$, $\theta = 10$.

    a. $P(X > 5) = 1 - f(5) = 1 - \left(1 - e^{-(5)^2/10}\right) = e^{-2.5} = 0.08208$

    b. Let Y = number of resistors out of three that fail before 5,000 hours. Then Y has a binomial distribution with parameters $n = 3$, $p = P(X < 5) = F(5)$ $= 1 - e^{-2.5}$, and

$$P(Y = 1) = \binom{3}{1}(1 - e^{-2.5})^1(e^{-2.5})^2 = 0.01855.$$

c. $E(X) = \theta^{1/\gamma}\Gamma\left(1+\dfrac{1}{\gamma}\right) = 10^{1/2}\Gamma\left(1+\dfrac{1}{2}\right) = 10^{1/2}\left(\dfrac{1}{2}\right)\Gamma\left(\dfrac{1}{2}\right) = \dfrac{\sqrt{10\pi}}{2}$

$\quad = 2.8025$

$V(X) = \theta^{2/\gamma}\left\{\Gamma\left(1+\dfrac{2}{\gamma}\right) - \left[\Gamma\left(1+\dfrac{1}{\gamma}\right)\right]^2\right\} = 10^{2/2}\left\{\Gamma\left(1+\dfrac{2}{2}\right) - \left[\Gamma\left(1+\dfrac{1}{2}\right)\right]^2\right\}$

$\quad = 10\left\{1 - \left[\dfrac{1}{2}\Gamma\left(\dfrac{1}{2}\right)\right]^2\right\} = 10\left(1-\dfrac{\pi}{4}\right) = 2.1460$

4.89   Let X = maximum wind-gust velocity. Then X has a Weibull distribution with parameters $\gamma = 2$, $\theta = 400$. Let k = the desired velocity such that $0.01 = P(X > k) = 1 - F(k) = 1 - \left(1 - e^{-(k)^2/400}\right) = e^{-(k)^2/400}$. Then, solving for k, we have $k = \sqrt{400[-\ln(0.01)]} = 42.9193$ feet per second.

4.91 $R(t_2 + t_1|t_1) = P(X > t_2 + t_1|X > t_1)$

$\quad = \dfrac{P(X > t_2 + t_1, X > t_1)}{P(X > t_1)}$

$\quad = \dfrac{P(X > t_2 + t_1)}{P(X > t_1)}$

$\quad = \dfrac{e^{-(t_2+t_1)/\theta}}{e^{-t_1/\theta}}$

$\quad = e^{-t_2/\theta}, t_2 > 0$

$\quad = R(t_2)$

4.93   For system (a), the reliability would be

$\quad R_s^a(t) = 1 - [1 - (R(t))^2][1 - R(t)]$

$\quad = R(t) + (R(t))^2 - (R(t))^3.$

For system (b), the reliability would be

$\quad R_s^b(t) = R(t)[1 - [1 - R(t)]^2]$

$\quad = 2(R(t))^2 - (R(t))^3.$

Consider $R_s^a(t) - R_s^b(t) = R(t) + (R(t))^2 - (R(t))^3 - 2(R(t))^2 + (R(t))^3$

$= R(t) - (R(t))^2 > 0.$

So system (a) is more reliable.

4.95 $M(t) = E(e^{tX}) = \int_0^\infty e^{tx} \dfrac{1}{\Gamma(\alpha)\beta^\alpha} x^{(\alpha-1)} e^{-x/\beta} dx$

$= \int_0^\infty \dfrac{1}{\Gamma(\alpha)\beta^\alpha} x^{(\alpha-1)} e^{-x(1/\beta-t)} dx \left(t < \dfrac{1}{\beta} \right)$

$= (1-\beta t)^{-\alpha} \int_0^\infty \dfrac{1}{\Gamma(\alpha) \left(\dfrac{\beta}{1-\beta t} \right)^\alpha} x^{(\alpha-1)} e^{-x/(\beta/(1-\beta t))} dx = (1-\beta t)^{-\alpha}$

4.97 $M_Z(t) = E(e^{tZ}) = \int_{-\infty}^\infty e^{tZ} \dfrac{1}{\sqrt{2\pi}} e^{-x^2/2} dz = e^{t^2/2} \int_{-\infty}^\infty \dfrac{1}{\sqrt{2\pi}} e^{-(z-t)^2/2} dz$

Since the integrand is a normal density function with parameters $\mu = t$,

$\sigma = 1$, we have $M_Z(t) = e^{t^2/2}$.

4.99

a. $1 = \int_{-\infty}^\infty f(y) dy = \int_0^2 cy\, dy = c \left(\dfrac{y^2}{2} \right) \Big|_0^2 = 2c.$ Therefore, $c = \dfrac{1}{2}.$

b. $F(y) = \begin{cases} 0 & y < 0 \\ \int_0^y \dfrac{x}{2} dx = \dfrac{x^2}{4} \Big|_0^y = \dfrac{y^2}{4} & 0 \le y \le 2 \\ 1 & y > 2 \end{cases}$

c.

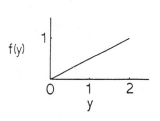

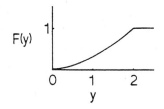

d. $P(1 \leq Y \leq 2) = F(2) - F(1) = \dfrac{2^2}{4} - \dfrac{1^2}{4} = \dfrac{3}{4}$

e. $P(1 \leq Y \leq 2) = \text{area of trapezoid} = \dfrac{1}{2}[f(2) + f(1)] \cdot (2 - 1) = \dfrac{1}{2}\left(1 + \dfrac{1}{2}\right) \cdot 1$

$= \dfrac{3}{4}$

4.101

a. $1 = \int_{-\infty}^{\infty} f(y)\,dy = \int_{-1}^{0} 0.2\,dy + \int_{0}^{1}(0.2 + cy)\,dy = 0.2y \Big|_{-1}^{0} + \left(0.2y + c\dfrac{y^2}{2}\right)\Big|_{0}^{1}$

$= 0.4 + \dfrac{c}{2}$

Therefore, $c = \dfrac{6}{5}$.

b. $F(y) = \begin{cases} 0 & y \leq -1 \\ \int_{-1}^{0} 0.2\,dx = 0.2(y + 1) & -1 < y \leq 0 \\ \int_{-1}^{0} 0.2\,dx + \int_{0}^{y}\left(0.2 + \dfrac{6}{5}x\right)dx = 0.2 + 0.2y + \dfrac{3}{5}y^2 & 0 < y \leq 1 \\ 1 & y > 1 \end{cases}$

c.

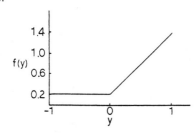

d. $F(-1) = 0$

$F(0) = 0.2(0 + 1) = 0.2$

$F(1) = 1$

e. $P(0 \leq Y \leq 0.5) = F(0.5) - F(0) = \left[0.2 + 0.2\left(\dfrac{1}{2}\right) + \dfrac{3}{5}\left(\dfrac{1}{2}\right)^2\right] - 0.2(0 + 1)$

$= \dfrac{1}{4}$

f. $E(Y) = \int_{-1}^{0} \frac{y}{5} dy + \int_{0}^{1} \left(\frac{y}{5} + \frac{6y^2}{5} \right) dy = \left. \frac{y^2}{10} \right|_{-1}^{0} + \left. \left(\frac{y^2}{10} + \frac{2}{5} y^3 \right) \right|_{0}^{1} = \frac{2}{5}$

$E(Y^2) = \int_{-1}^{0} \frac{y^2}{5} dy + \int_{0}^{1} \left(\frac{y^2}{5} + \frac{6y^3}{5} \right) dy = \left. \frac{y^3}{15} \right|_{-1}^{0} + \left. \left(\frac{y^3}{15} + \frac{3}{10} y^4 \right) \right|_{0}^{1} = \frac{13}{30}$

$V(X) = E(X^2) - [E(X)]^2 = \frac{13}{30} - \frac{4}{25} = \frac{41}{150}$

4.103 $P(X < 1.9) = P \left(\frac{X - 2.4}{0.5} < \frac{1.9 - 2.4}{0.5} \right) = P(Z < -1) = 0.5 - .3413 = 0.1587$

i.e., 15.87% of the students will be dropped.

4.105   Let X = bearing diameter. Then

P(bearing is scrap) $= P(X > 3.002) + P(X < 2.998)$

$= P \left(\frac{X - 3.0005}{0.001} > \frac{3.002 - 3.0005}{0.001} \right) + P \left(\frac{X - 3.0005}{0.001} < \frac{2.998 - 3.0005}{0.001} \right)$

$= P(Z > 1.5) + P(Z < 2.50) = 0.0668 + 0.0062 = 0.073.$

4.107

a. $E(Y) = \int_{-\infty}^{\infty} y f(y) dy = \int_{0}^{\infty} cy^2 e^{-2y} dy = c \left(\frac{1}{2} \right)^3 \Gamma(3) = \frac{c}{4}$ since $\int_{0}^{\infty} x^{n-1} e^{-\frac{x}{\theta}} dx$

$= \Gamma(n) \theta^n$

$E(Y^2) = \int_{-\infty}^{\infty} y^2 f(y) dy = \int_{0}^{\infty} cy^3 e^{-2y} dy = c \left(\frac{1}{2} \right)^4 \Gamma(4) = \frac{3}{8} c$

$V(Y) = E(Y^2) - [E(Y)]^2 = \frac{3}{8} c - \frac{c^2}{16} = \frac{c(6 - c)}{16}$

b. $M(t) = E(e^{tY}) = \int_{-\infty}^{\infty} e^{ty} f(y) dy = \int_{0}^{\infty} ce^{ty} ye^{-2y} dy = \int_{0}^{\infty} cye^{-(2-t)y} dy$

$= c\Gamma(2) \left(\frac{1}{2 - t} \right)^2 = \frac{c}{(2 - t)^2}$

c. $1 = \int_{-\infty}^{\infty} f(y) dy = \int_{0}^{\infty} cye^{-2y} dy = c\Gamma(2) \left(\frac{1}{2} \right)^2 = \frac{c}{4}.$ Therefore, $c = 4.$

4.109 Let X = yield force of the steel reinforcing bar. Then X has a normal distribution with parameters $\mu = 8,500$, $\sigma = 80$. Let Y = number of bars out of three used that have yield forces in excess of 8,700. Then Y has a binomial distribution with parameters $n = 3$, $p = P(X > 8,700) = P\left(\dfrac{X - 8,500}{80} > \dfrac{8,700 - 8,500}{80}\right) =$

$P(Z > 2.5) = 0.5 - 0.4938 = 0.0062$. Then

$$P(Y = 3) = \binom{3}{3}(0.0062)^3(1 - 0.0062)^0 = (0.0062)^3 = 2.3833 \times 10^{-7}$$

4.111 Let X = gap time. Then X has an exponential distribution with parameters $\theta = 10$, which is equivalent to a Gamma distribution with parameters $\alpha = 1$, $\beta = \theta = 10$.

a. $P(X \le 60) = F(60) = 1 - e^{60/10} = 1 - e^{-6} = 0.9975$.

b. Assuming independent gap times, Y, the sum of the next four gap times, has a Gamma distribution with parameters $\sigma = 4$, $\beta = 10$; i.e.,

$$f(y) = \begin{cases} \dfrac{1}{\Gamma(4)(10)^4}y^{(4-1)}e^{-(y/10)} = \dfrac{1}{60,000}y^3 e^{-(y/10)} & y > 0 \\ 0 & y \le 0. \end{cases}$$

4.113 $P(X < 4 | X \ge 2) = \dfrac{P(2 \le X \le 4)}{P(X \ge 2)} = \dfrac{F(4) - F(2)}{1 - F(2)} = \dfrac{1 - e^{-4/3} - (1 - e^{2^2/3})}{1 - (1 - e^{-2^2/3})}$

$= \dfrac{e^{-4/3} - e^{-16/3}}{e^{-4/3}} = 1 - e^{-4} = 0.9817$

4.115 Let c = budgeted amount in hundreds of dollars.

$$0.10 = P(Y > c) = \int_c^1 3(1-y)^2 dy = -(1-y)^3 \Big|_c^1 = (1-c)^3$$

i.e., $c = 1 - \left(\dfrac{1}{10}\right)^{1/3} = 0.5358$, i.e., the budgeted amount should be \$53.58.

4.117 For Y a gamma random variable with parameters $\beta = 1$, α, and X a Poisson random variable with parameter λ, the given relationship may be written as $P(Y > \lambda) = P(X \le \alpha - 1)$. Thus, for $\alpha = 2, \beta = 1, \lambda = 1$,

$P(Y > 1) = P(X \le 1) = 0.736$

4.119

    a. Let X = number of plants in a circular region of radius r. Then X has a Poisson distribution with parameters $\lambda \pi r^2$. Thus

$$P(R > r) = P(\text{no plants in an area of } \pi r^2) = P(X = 0) = \frac{(\lambda \pi r^2)^0 e^{\lambda \pi r^2}}{0!}$$

$$= e^{\lambda \pi r^2}$$

    Therefore, $F(r) = 1 - P(R > r) = 1 - e^{\lambda \pi r^2}$ so that R has a Weibull distribution with parameters $\gamma = 2, \theta = \dfrac{1}{\lambda \pi}$.

    b. $E(R) = \theta^{1/\gamma} \Gamma\left(1 + \dfrac{1}{\gamma}\right) = (\lambda \pi)^{-1/2} \Gamma\left(1 + \dfrac{1}{2}\right) = (\lambda \pi)^{-1/2} \dfrac{1}{2} \pi^{1/2} = \dfrac{1}{2\sqrt{\lambda}}$

4.121   Let X = gram weight of the certain type of aluminum. Then X has a lognormal distribution with parameters $\mu = 3, \sigma = 4$.

    a. $E(X) = e^{\mu + \frac{\sigma^2}{2}} = e^{3 + \frac{4^2}{2}} = e^{11} = 59874.142$ in units of 10^{-2} grams

    $V(X) = e^{2\mu + \sigma^2}(e^{\sigma^2} - 1) = e^{2(3) + 4^2}(e^{4^2} - 1) = e^{38} - e^{22}$ in units of $10^{-4} g^2$

    b. By Tchebysheff's theorem, with $k = 2$, the required interval, in grams, is

$$[E(X) - 2\sqrt{V(X)}, E(X) + 2\sqrt{V(X)}]$$

$$= [e^{11} \times 10^{-2} - 2\sqrt{(e^{38} - e^{22}) \times 10^{-4}}, e^{11} \times 10^{-2} + 2\sqrt{(e^{38} - e^{22}) \times 10^{-4}}]$$

$$= (-3569047.08, 3570244.56).$$

    Since gram weights are nonnegative, the interval becomes $(0, 3570244.56)$.

    c. $P(X < 598.74) = P(Y < 6.3948) = P\left[Z < \dfrac{6.3948 - 3}{4}\right]$

    $= P(Z < 0.8487) = 0.8023$

Chapter 5.

5.1 a., b.

Firms

Contract 1	Contract 2	Probability	$(x_1,\quad x_2)$
I	I	$\frac{1}{9}$	(2,0)
I	II	$\frac{1}{9}$	(1,1)
I	III	$\frac{1}{9}$	(1,0)
II	I	$\frac{1}{9}$	(1,1)
II	II	$\frac{1}{9}$	(0,2)
II	III	$\frac{1}{9}$	(0,1)
III	I	$\frac{1}{9}$	(1,0)
III	II	$\frac{1}{9}$	(0,1)
III	III	$\frac{1}{9}$	(0,0)

		x_1		
		0	1	2
	0	$\frac{1}{9}$	$\frac{2}{9}$	$\frac{1}{9}$
x_2	1	$\frac{2}{9}$	$\frac{2}{9}$	0
	2	$\frac{1}{9}$	0	0
$p(x_1)$		$\frac{4}{9}$	$\frac{4}{9}$	$\frac{1}{9}$

c. $P(X_1 = 1 | X_2 = 1) = \dfrac{P(X_1 = 1, \quad X_2 = 1)}{P(X_2 = 1)} = \dfrac{\frac{2}{9}}{\frac{2}{9} + \frac{2}{9}} = \dfrac{1}{2}$

5.3

a.

	X_1	0	1
		0	1
	0	0.0635	0.0775
	1	0.1007	0.0556
X_2	2	0.1630	0.0653
	3	0.1691	0.0549
	4	0.1929	0.0574

b.

X_1	$P(X_1|X_2 = 0)$	$P(X_1|X_2 = 1)$	$P(X_1|X_2 = 2)$
0	0.4502	0.6455	0.7139
1	0.5498	0.3558	0.2861

X_1	$P(X_1|X_2 = 3)$	$P(X_1|X_2 = 4)$
0	0.7548	0.7707
1	0.2452	0.2293

Obviously, the older the child is, the better his/her chance of survival in a car accident without wearing a seat belt.

c.

X_2	$P(X_2|X_1 = 0)$	$P(X_2|X_1 = 1)$
0	0.0921	0.2495
1	0.1461	0.1788
2	0.2365	0.2102
3	0.2453	0.1768
4	0.2799	0.1847

No, this implies that if a child survives, then s/he will probably be older.

5.5

a. $f_1(x_1) = \begin{cases} \int_0^1 1 dx_2 = 1 & 0 \leq x_1 \leq 1 \\ 0 & \text{otherwise} \end{cases}$

b. $P\left(X_1 \leq \dfrac{1}{2}\right) = \int_0^{1/2} 1 dx_1 = \dfrac{1}{2}$

c. $f_1(x_1) f_2(x_2) = 1 = f(x_1 x_2)$, for $0 \leq x_1 \leq 1$, $\quad 0 \leq x_2 \leq 1$. Therefore, X_1 and X_2 are independent.

5.7

a. $P\left(X_1 \leq \dfrac{3}{4}, \quad X_2 \leq \dfrac{3}{4}\right) = \int_0^{1/4} \int_0^{3/4} 2 dx_2 dx_1 + \int_{1/4}^{3/4} \int_0^{1-x_1} 2 dx_2 dx_1$

$= 2 \left(\dfrac{1}{4}\right)\left(\dfrac{3}{4}\right) + \int_{1/4}^{3/4} 2(1 - x_1) dx_1 = \dfrac{3}{8} + \dfrac{1}{2} = \dfrac{7}{8}$

b. $P\left(X_1 \leq \dfrac{1}{2}, \quad X_2 \leq \dfrac{1}{2}\right) = 2 \left(\dfrac{1}{2}\right)\left(\dfrac{1}{2}\right) = \dfrac{1}{2}$

c. $P\left(X_1 \leq \dfrac{1}{2} | X_2 \leq \dfrac{1}{2}\right) = \dfrac{P\left(X_1 \leq \dfrac{1}{2}, \quad X_2 \leq \dfrac{1}{2}\right)}{P\left(X_2 \leq \dfrac{1}{2}\right)} = \dfrac{\dfrac{1}{2}}{\int_0^{1/2} \int_0^{1-x_2} 2 dx_1 dx_2}$

$= \dfrac{1}{2 \int_0^{1/2} 2(1 - x_2) dx_2} = \dfrac{2}{3}$

5.9

a. $P\left(X_1 < \dfrac{1}{2}, \quad X_2 > \dfrac{1}{4}\right) = \int_0^{1/2} \int_{1/4}^1 (x_1 + x_2) dx_2 dx_1 = \int_0^{1/2}\left(\dfrac{3x_1}{4} + \dfrac{15}{32}\right) dx_1$

$= \dfrac{21}{64}$

b. $P(X_1 + X_2 \leq 1) = \int_0^1 \int_0^{1-x_1} (x_1 + x_2) dx_2 dx_1 = \int_0^1 \left.\left(x_1 x_2 + \dfrac{x_2^2}{2}\right)\right|_0^{1-x_1} dx_1$

$= \int_0^1 \dfrac{1}{2}(1 - x_1^2) dx_1 = \dfrac{1}{3}$

c. Note that

$$f_1(x_1)f_2(x_2) = \left(x_1 + \frac{1}{2}\right)\left(x_2 + \frac{1}{2}\right) \neq x_1 + x_2 = f(x_1, x_2)$$

for $0 \leq x_1 \leq 1$, $\quad 0 \leq x_2 \leq 1$.

Therefore, X_1 and X_2 are not independent.

5.11   Note that

$$f_1(x_1) = \int_0^\infty \frac{1}{8}x_1 e^{-x_1/2} e^{-x_2/2} dx_2 = \frac{1}{4}x_1 e^{-x_1/2}\left(-e^{-x_2/2}\right)\Big|_0^\infty$$

$$= \frac{1}{4}x_1 e^{-x_1/2} \text{ for } 0 < x_1 < \infty$$

and, integrating by parts, we have

$$f_2(x_2) = \int_0^\infty \frac{1}{8}x_1 e^{-x_1/2} e^{-x_2/2} dx_1 = \frac{1}{8}e^{-x_2/2}\left\{-2x_1 e^{-x_1/2}\Big|_0^\infty + 2\int_0^\infty e^{-x_1/2}dx_1\right\}$$

$$= \frac{1}{8}e^{-x_2/2}\left\{-4e^{-x_1/2}\Big|_0^\infty\right\} = \frac{1}{2}e^{-x_2/2} \text{ for } 0 < x_2 < \infty.$$

a. Note that

$$f_1(x_1)f_2(x_2) = \frac{1}{4}x_1 e^{-x_1/2}\frac{1}{2}e^{-x_2/2} = \frac{1}{8}x_1 e^{-(x_1+x_2)/2} = f(x_1, x_2)$$

for $0 < x_1 < \infty$, $\quad 0 < x_2 < \infty$.

Therefore, X_1 and X_2 are independent.

b. $P(X_1 > 1, \quad X_2 > 1) = P(X_1 > 1)P(X_2 > 1) = \int_1^\infty \frac{1}{4}x_1 e^{-x_1/2}dx_1 \int_1^\infty \frac{1}{2}e^{-x_2/2}dx_2$

$$= \left\{-\frac{1}{2}x_1 e^{-x_1/2}\Big|_1^\infty + \frac{1}{2}\int_1^\infty e^{-x_1/2}dx_1\right\}\left\{-e^{-x_2/2}\Big|_1^\infty\right\}$$

$$= \left\{\frac{1}{2}e^{-1/2} + e^{-1/2}\right\}e^{-1/2} = \frac{3e^{-1}}{2} = 0.5518$$

5.13   Let $X_i =$ arrival time for friend i, $0 \leq x_i \leq 1$, $i = 1, 2$. The two friends will meet

if $|x_1 - x_2| < \frac{1}{3}$ and

$$P\left(|X_1 - X_2| < \frac{1}{3}\right) = \int_0^{1/6} \int_0^{x_1+(1/6)} 1 dx_2 dx_1 + \int_{1/6}^{5/6} \int_{x_1-(1/6)}^{x_1+(1/6)} 1 dx_2 dx_1$$

$$+ \int_{5/6}^1 \int_{x_1-(1/6)}^1 1 dx_2 dx_1 = \frac{3}{72} + \frac{2}{9} + \frac{3}{72} = \frac{11}{36}.$$

5.15 Let $X_i =$ the time the ith call is made, $i = 1, 2, 0 \le x_i \le 1$. Then

$$f(x_1, \quad x_2) = f_1(x_1)f_2(x_2) = \begin{cases} 1 & 0 \le x_1 \le 1, \quad 0 \le x_2 \le 1 \\ 0 & \text{otherwise} \end{cases}$$

a. $P\left(X_1 < \dfrac{1}{2}, X_2 < \dfrac{1}{2}\right) = P\left(X_1 < \dfrac{1}{2}\right) P\left(X_2 < \dfrac{1}{2}\right) = \left(\dfrac{1}{2}\right)\left(\dfrac{1}{2}\right) = \dfrac{1}{4}$

b. $P\left(|X_1 - X_2| < \dfrac{1}{12}\right) = \int_0^{1/12} \int_0^{x_1+(1/12)} dx_2 dx_1 + \int_{1/12}^{11/12} \int_{x_1-(1/12)}^{x_1+(1/12)} dx_2 dx_1$

$+ \int_{11/12}^{1} \int_{x_1-(1/12)}^{1} dx_2 dx_1 = \dfrac{3}{288} + \dfrac{20}{144} + \dfrac{3}{288} = \dfrac{23}{144}$

5.17

a. $E(X_1) = (0)(0.8) + (1)(0.2) = 0.2$

$E(X_1^2) = (0)^2(0.8) + (1)^2(0.2) = 0.2$

$V(X_1) = E(X_1^2) - [E(X_1)]^2 = 0.2 - (0.2)^2 = 0.16$

Note that X_2 has the same marginal distribution as X_1, and so its mean and variance are equal to those of X_1.

b. Cov $(X_1, \quad X_2) = E(X_1 X_2) - E(X_1)E(X_2) = 0 - (0.2)^2 = -0.04$

c. $E(Y) = E(X_1 + X_2) = E(X_1) + E(X_2) = 0.2 + 0.2 = 0.4$

$V(Y) = V(X_1 + X_2) = V(X_1) + V(X_2) + 2\text{Cov}(X_1, \quad X_2) = 0.16 + 0.16$

$+ 2(-0.04) = 0.24$

5.19

a. $E(X_1 + X_2) = \int_0^1 \int_0^{1-x_1} (x_1 + x_2) 2 dx_2 dx_1 = \int_0^1 (1 - x_1^2) dx_1 = \dfrac{2}{3}$

$E[(X_1+X_2)^2] = \int_0^1 \int_0^{1-x_1} (x_1+x_2)^2 2 dx_2 dx_1 = \int_0^1 \dfrac{2}{3}\left(1 - x_1^3\right) dx_1 = \dfrac{2}{3}\left(x - \dfrac{x^4}{4}\right)\Big|_0^1$

$= \dfrac{2}{3}\left(\dfrac{3}{4}\right) = \dfrac{1}{2}$

$V(Y) = E(Y^2) - [E(Y)]^2 = \dfrac{1}{2} - \left(\dfrac{2}{3}\right)^2 = \dfrac{1}{18}$

b. Using Tchebysheff's theorem with $k = \sqrt{2}$, we have

$$P\left(\frac{2}{3} - \sqrt{\frac{2}{18}} < X_1 + X_2 < \frac{2}{3} + \sqrt{\frac{2}{18}}\right) \geq 0.5; \text{ i.e., the desired interval is}$$

$$\left(\frac{1}{3}, \ 1\right).$$

c. $\rho = \dfrac{E(X_1 X_2) - E(X_1)E(X_2)}{\sqrt{\text{Var}(X_1)\text{Var}(X_2)}}$

$f(X_1) = \int_0^{1-X_1} f(X_1, \quad X_2)dx_2 = \int_0^{1-X_1} 2dX_2 = 2(1 - X_1), 0 \leq X_1 \leq 1$

$f(X_2) = \int_0^{1-X_2} f(X_1, \quad X_2)dX_1 = \int_0^{1-X_1} 2dX_1 = 2(1 - X_2), 0 \leq X_2 \leq 1$

So $E(X_1) = E(X_2) = \int_0^1 y2(1 - y)dy = 2\int_0^1 y - y^2 dy$

$$= 2\left(\frac{y^2}{2} - \frac{y^3}{3}\right)\bigg|_0^1 = 2\left(\frac{1}{2} - \frac{1}{3}\right) = 2\left(\frac{1}{6}\right) = \frac{1}{3}$$

and $E(X_1^2) = E(X_2^2) = \int_0^1 y^2 2(1 - y)dy = 2\int_0^1 y^2 - y^3 dy$

$$= 2\left(\frac{y^3}{3} - \frac{y^4}{4}\right)\bigg|_0^1 = 2\left(\frac{1}{3} - \frac{1}{4}\right) = 2\left(\frac{1}{12}\right) = \frac{1}{6}.$$

Hence, Var $(X_1) = \text{Var}(X_2) = E(X_1^2) - (E(X_1))^2 = \frac{1}{6} - \left(\frac{1}{3}\right)^2$

$$= \frac{1}{6} - \frac{1}{9} = \frac{1}{18}$$

$E(X_1 X_2) = \int_0^1 \int_0^{1-x_1} x_1 x_2 2 dx_1 dx_2 = \int_0^1 x_1 \int_0^{1-x_1} 2x_2 dx_2 dx_1$

$$= \int_0^1 x_1\left(x_2^2\bigg|_0^{1-x_1}\right)dx_1 = \int_0^1 x_1(1 - x_1)^2 dx_1$$

$$= \int_0^1 x_1 - 2x_1^2 + x_1^3 dx_1 = \frac{x_1^2}{2} - \frac{2x_1^3}{3} + \frac{x_1^4}{4}\bigg|_0^1$$

$$= \frac{1}{2} - \frac{2}{3} + \frac{1}{4} = \frac{1}{12}$$

So $\rho = \dfrac{\dfrac{1}{12} - \left(\dfrac{1}{3}\right)^2}{\sqrt{\left(\dfrac{1}{18}\right)^2}} = \dfrac{\dfrac{1}{12} - \dfrac{1}{9}}{\dfrac{1}{18}} = \dfrac{\dfrac{3-4}{36}}{\dfrac{1}{18}} = -\dfrac{1}{2}$

5.21

   a. $P(Y_1 < 2, \quad Y_2 > 1) = \int_1^2 \int_{y_2}^2 e^{-y_1} dy_1 dy_2 = \int_1^2 (e^{-y_2} - e^{-2}) dy_2 = e^{-1} - 2e^{-2}$

      $= 0.0972$

   b. $P(Y_1 > 2Y_2) = \int_0^\infty \int_{2y_2}^\infty e^{-y_1} dy_1 dy_2 = \int_0^\infty e^{-2y_2} dy_2 = \dfrac{1}{2}$

   c. $P(Y_1 - Y_2 \geq 1) = \int_0^\infty \int_{1+y_2}^\infty e^{-y_1} dy_1 dy_2 = \int_0^\infty e^{-(1+y_2)} dy_2 = e^{-1}$

   d. $f_1(y_1) = \begin{cases} \int_0^{y_1} e^{-y_1} dy_2 = y_1 e^{-y_1} & 0 \leq y_1 < \infty \\ 0 & \text{otherwise} \end{cases}$

      $f_2(y_2) = \begin{cases} \int_{y_2}^\infty e^{-y_1} dy_1 = e^{-y_2} & 0 \leq y_2 < \infty \\ 0 & \text{otherwise} \end{cases}$

5.23

   a. $E(Y_1 - Y_2) = \int_0^\infty \int_{y_2}^\infty (y_1 - y_2) e^{-y_1} dy_1 dy_2 = \int_0^\infty e^{-y_2} dy_2 = 1$

   b. $E[(Y_1 - Y_2)^2] = \int_0^\infty \int_{y_2}^\infty (y_1 - y_2)^2 e^{-y_1} dy_1 dy_2 = \int_0^\infty 2e^{-y_2} dy_2 = 2$

      $V(Y) = E(Y^2) - [E(Y)]^2 = 2 - 1^2 = 1$

   c. No, since

      $P(Y_1 - Y_2 > 2) = \int_0^\infty \int_{2+y_2}^\infty e^{-y_1} dy_1 dy_2 = \int_0^\infty e^{-(2+y_2)} dy_2 = e^{-2} = 0.1353.$

5.25 $\text{Cov}(X_1, \quad X_2) = E[(X_1 - \mu_1)(X_2 - \mu_2)]$

      $= \int_{-\infty}^\infty \int_{-\infty}^\infty (x_1 - \mu_1)(x_2 - \mu_2) f(x_1, x_2) dx_1 dx_2$

      $= \int_{-\infty}^\infty \int_{-\infty}^\infty (x_1 x_2 - x_1 \mu_2 - x_2 \mu_1 + \mu_1 \mu_2) f(x_1, x_2) dx_1 dx_2$

      $= \int_{-\infty}^\infty \int_{-\infty}^\infty (x_1 x_2) f(x_1 x_2) dx_1 dx_2 - \mu_2 \int_{-\infty}^\infty \int_{-\infty}^\infty x_1 f(x_1 x_2) dx_1 dx_2$

      $- \quad \mu_2 \int_{-\infty}^\infty \int_{-\infty}^\infty x_2 f(x_1 x_2) dx_1 dx_2 + \mu_1 \mu_2 \int_{-\infty}^\infty \int_{-\infty}^\infty f(x_1 x_2) dx_1 dx_2$

      $= E(X_1 X_2) - \mu_1 \mu_2 - \mu_1 \mu_2 + \mu_1 \mu_2 = E(X_1 X_2) - \mu_1 \mu_2$

5.27 Let $C = 20,000Y_1 + 10,000Y_2 + 2,000Y_3$, where $Y_1,\quad Y_2,\quad Y_3$ are as defined in Exercise 5.26. Then

$$E(C) = 20,000E(Y_1) + 10,000E(Y_2) + 2,000E(Y_3) = 20,000np_1 + 10,000np_2$$
$$+2,000np_3$$
$$= 20,000(4)(0.73) + 10,000(4)(0.20) + 2,000(4)(0.07) = 66,960.$$

5.29 Let $Y_1 =$ number of days having total radiation of at most 5 calories, $Y_2 =$ number of days having total radiation between 5 and 6 calories, and $Y_3 =$ number of days having radiation between 6 and 8 calories. Then $(Y_1,\quad Y_2,\quad Y_3)$ has a multinomial distribution with parameters $n = 6$, $p_1 = 0.30,\quad p_2 = 0.60 - 0.30 = 0.30,\quad p_3 = 1 - 0.60 = 0.40$. Therefore,

$$P(Y_1 = 3,\quad Y_2 = 1,\quad Y_3 = 2) = \frac{6!}{3!1!2!}(0.30)^3(0.30)^1(0.40)^2 = 0.07776.$$

This solution is based on the assumption that measurements on different days are independent.

5.31 Let $Y_1 =$ number of customers passing through gate A, $Y_2 =$ number of customers passing through gate B, and $Y_3 =$ number of customers passing through gate C. Then $(Y_1,\quad Y_2,\quad Y_3)$ has a multinomial distribution with parameters n = 4, $p_1 = p_2 = p_3 = \dfrac{1}{3}$.

a. $P(Y_1 = 2,\quad Y_2 = 1,\quad Y_3 = 1) = \dfrac{4!}{2!1!1!}\left(\dfrac{1}{3}\right)^2\left(\dfrac{1}{3}\right)^2\left(\dfrac{1}{3}\right)^1 = \dfrac{4}{27}$

b. P(all four select same gate)

$$= P(Y_1 = 4,\quad Y_2 = 0,\quad Y_3 = 0) + P(Y_1 = 0,\quad Y_2 = 4,\quad Y_3 = 0)$$
$$+P(Y_1 = 0,\quad Y_2 = 0,\quad Y_3 = 4)$$
$$= (3)\dfrac{4!}{4!0!0!}\left(\dfrac{1}{3}\right)^4\left(\dfrac{1}{3}\right)^0\left(\dfrac{1}{3}\right)^0 = \dfrac{1}{27}$$

c. P(all three gates are used)

$$= P(Y_1 = 2, \quad Y_2 = 1, \quad Y_3 = 1) + P(Y_1 = 1, \quad Y_2 = 2, \quad Y_3 = 1)$$

$$+ P(Y_1 = 1, \quad Y_2 = 1, \quad Y_3 = 2)$$

$$= (3)\frac{4!}{1!2!1!}\left(\frac{1}{3}\right)^1\left(\frac{1}{3}\right)^2\left(\frac{1}{3}\right)^1 = \frac{4}{9}$$

5.33 Let Y_0 = number out of the selected items with no defects. Then $(Y_0, \quad Y_1, \quad Y_2)$ has a multinomial distribution with parameters $n = 10, p_0 = 0.85, p_1 = 0.10, p_2 = 0.05$, and $E(Y_i) = np_i, V(Y_i) = np_i(1 - p_i)$, $\mathrm{Cov}(Y_i, \quad Y_j) = -np_i p_j, \quad i, j = 1, 2, 3, i \neq j$. Therefore we have

$$E(Y_1 + 3Y_2) = E(Y_1) + 3E(Y_2) = 10(0.10) + 3(10)(0.05) = 2.5$$

and

$$V(Y_1 + 3Y_2) = V(Y_1) + (3)^2 V(Y_2) + 2(3)\mathrm{Cov}(Y_1, \quad Y_2)$$

$$= 10(0.10)(0.90) + 9(10)(0.05)(0.95) - 2(3)(10)(0.10)(0.05) = 4.875.$$

5.35 Let Y_1 = number of vehicles that turn left, Y_2 = number of vehicles that turn right, and Y_3 = number of vehicles that continue straight, out of n vehicles arriving at the intersection. Then $(Y_1, \quad Y_2, \quad Y_3)$ has a multinomial distribution with parameters n, $p_1 = 0.4, p_2 = 0.25, p_3 = 0.35$.

a. $P(Y_1 = 1, \quad Y_2 = 1, \quad Y_3 = 3) = \dfrac{5!}{1!1!3!}(0.4)(0.25)(0.35)^3 = 0.08575$

b. Note that the marginal distribution of Y_2 is binomial with parameters $n = 5$, $p = 0.25$, so

$$P(Y_2 \geq 1) = 1 - P(Y_2 = 0) = 1 - \binom{5}{0}(0.25)^0(0.75)^5 = 0.7627.$$

c. Note that, assuming that vehicles turn independently of each other, the marginal distribution of Y_1 is binomial with parameters $n = 100$, $p = 0.4$, so

$$E(Y_1) = np = 100(0.4) = 40$$

$$V(Y_1) = np(1 - p) = 100(0.4)(0.6) = 24.$$

5.37 Let $X_1 = $ number of cars arriving at entrance I, $X_2 = $ number of cars arriving at entrance II, and $Y = X_1 + X_2$. Then X_1 has a Poisson distribution with parameter $\lambda_1 = 3$, X_2 has a Poisson distribution with parameter $\lambda_2 = 4$, and Y has a Poisson distribution with parameter $\lambda_Y = \lambda_1 + \lambda_2 = 3 + 4 = 7$, and

$$P(Y = 3) = \frac{\lambda_Y^3 e^{\lambda_Y}}{3!} = \frac{7^3 e^{-7}}{6} = 0.05213.$$

5.39 Let $X = $ resistance of resistor. Then X has a normal distribution with parameters $\mu_X = 100$, $\sigma_X^2 = 100$, and, from Exercise 5.38, $Y = 2X$ has a normal distribution with parameters $\mu_Y = 2\mu_X = 2(100) = 200$, and $\sigma_Y^2 = 2^2 \sigma_X^2 = 4(100)$.

a. $P(Y > 220) = P\left(\dfrac{Y - 200}{\sqrt{400}} > \dfrac{220 - 200}{\sqrt{400}}\right) = P(Z > 1) = 0.5 - 0.3413$

$= 0.1587$

b. $P(Y < 190) = P\left(\dfrac{Y - 200}{\sqrt{400}} < \dfrac{190 - 200}{\sqrt{400}}\right) = P\left(Z < -\dfrac{1}{2}\right) = 0.5 - 0.1915$

$= 0.3085$

5.41

a. $1 = \int_{-\infty}^{\infty} \int_{-\infty}^{\infty} f(x_1, \quad x_2) dx_1 dx_2 = k \int_0^1 \int_0^1 x_1 x_2 dx_1 dx_2 = \dfrac{k}{2} \int_0^1 x_2 dx_2 = \dfrac{k}{4}.$

Hence, $k = 4$.

b. $f_1(x_1) = \int_0^1 f(x_1, \quad x_2) dx_2 = \int_0^1 4x_1 x_2 dx_2 = 2x_1$ for $0 \leq x_1 \leq 1$

Similarly, $f_2(x_2) = 2x_2$ for $0 \leq x_2 \leq 1$.

c. $F(x_1, \quad x_2) = \int_{-\infty}^{x_1} \int_{-\infty}^{x_2} f(y_1, \quad y_2) dy_1 dy_2$

$$= \begin{cases} 0 & x_1, \quad x_2 < 0 \\ \int_0^{x_1} \int_0^{x_2} 4y_1 y_2 dy_2 dy_2 = x_1^2 x_2^2 & 0 \leq x_1, \quad x_2 \leq 1 \\ \int_0^{x_1} \int_0^1 4y_1 y_2 dy_1 dy_2 = x_1^2 & 0 \leq x_1 \leq 1, \quad x_2 > 1 \\ x_2^2 & 0 \leq x_2 \leq 1, x_1 > 1 \\ 1 & x_1, \quad x_2 > 1 \end{cases}$$

d. $P\left(X_1 < \frac{1}{2},\ X_2 < \frac{3}{4}\right) = F\left(\frac{1}{2},\ \frac{3}{4}\right) = \left(\frac{1}{2}\right)^2 \left(\frac{3}{4}\right)^2 = \frac{9}{64}$

e. $P\left(X_1 \leq \frac{1}{2} | X_2 > \frac{3}{4}\right) = \dfrac{\int_0^{1/2} \int_{3/4}^1 x_1 x_2 dx_2 dx_1}{\int_{3/4}^1 2x_2 dx_2} = \dfrac{\left(1 - \frac{9}{16}\right)\left(\frac{1}{4}\right)}{\left(1 - \frac{9}{16}\right)} = \frac{1}{4}$

5.43

a. Note that

$$P\left(X_1 = x_1,\quad X_2 = x_2\right) = \frac{\binom{4}{x_1}\binom{3}{x_2}\binom{2}{3 - x_1 - x_2}}{\binom{9}{3}}$$

for $x_1 = 0, 1, 2, 3;\ x_2 = 0, 1, 2, 3;\ 1 \leq x_1 + x_2 \leq 3$; i.e., we have

x_1	x_2	$p(x_1,\quad x_2)$
0	0	0
0	1	$\frac{1}{28}$
0	2	$\frac{1}{14}$
0	3	$\frac{1}{84}$
1	0	$\frac{1}{21}$
1	1	$\frac{2}{7}$
1	2	$\frac{1}{7}$
2	0	$\frac{1}{7}$
2	1	$\frac{3}{14}$
3	0	$\frac{1}{21}$
otherwise		0

b.

			x_2			
		0	1	2	3	$p(x_1)$
	0	0	$\dfrac{1}{28}$	$\dfrac{1}{14}$	$\dfrac{1}{84}$	$\dfrac{10}{84}$
x_1	1	$\dfrac{1}{21}$	$\dfrac{2}{7}$	$\dfrac{1}{7}$	0	$\dfrac{10}{21}$
	2	$\dfrac{1}{7}$	$\dfrac{3}{14}$	0	0	$\dfrac{5}{14}$
	3	$\dfrac{1}{21}$	0	0	0	$\dfrac{1}{21}$
	$p(x_2)$	$\dfrac{5}{21}$	$\dfrac{15}{28}$	$\dfrac{3}{14}$	$\dfrac{1}{84}$	

c. $P(X_1 = 1 | X_2 \leq 1) = \dfrac{P(X_1 = 1, \ X_2 \leq 1)}{P(X_2 \geq 1)} = \dfrac{\frac{3}{7}}{\frac{16}{21}} = \dfrac{9}{16}$

5.45

a. $f(x_1 | x_2) = \dfrac{f(x_1, \ x_2)}{f_2(x_2)} = \dfrac{3x_1}{\frac{3}{2}(1 - x_2^2)} = \dfrac{2x_1}{1 - x_2^2}$ for $0 \leq x_2 \leq x_1 \leq 1$

b. $f(x_2 | x_1) = \dfrac{f(x_1, \ x_2)}{f_1(x_1)} = \dfrac{3x_1}{3x_1^2} = \dfrac{1}{x_1}$ for $0 \leq x_2 \leq x_1 \leq 1$

c. $f(x_1 | x_2) = \dfrac{f(x_1, \ x_2)}{f_2(x_2)} = \dfrac{2x_1}{1 - x_2^2} \neq f_1(x_1) = 3x_1^2$ for $0 \leq x_2 \leq x_1 \leq 1$

Therefore, X_1 and X_2 are dependent.

d. $P\left(X_1 \leq \dfrac{3}{4} \Big| X_2 = \dfrac{1}{2}\right) = \int_{-\infty}^{3/4} f\left(x_1 | x_2 = \dfrac{1}{2}\right) dx_1 = \int_{1/2}^{3/4} \dfrac{2x_1}{1 - \left(\frac{1}{2}\right)^2} dx_1$

$= \dfrac{4}{3}\left(\dfrac{9}{16} - \dfrac{1}{4}\right) = \dfrac{5}{12}$

5.47

a. $f(x_1) = \int_{-\infty}^{\infty} f(x_1, \quad x_2)dx_2$

$$= \begin{cases} \int_{-\sqrt{1-x_1^2}}^{\sqrt{1-x_2}} \frac{1}{\pi} dx_2 = \frac{2}{\pi}\sqrt{1-x_1^2} & -1 \le x_1 \le 1 \\ 0 & \text{otherwise} \end{cases}$$

b. In polar coordinates, we have

$$P(X_1 \le X_2) = \int_{\pi/4}^{5\pi/4} \int_2^1 \frac{1}{\pi} r\, dr\, d\theta = \int_{\pi/4}^{5\pi/4} \frac{1}{2\pi} d\theta = \frac{1}{2\pi}(\pi) = \frac{1}{2}.$$

5.49

a. $1 = \int_{-\infty}^{\infty} \int_{-\infty}^{\infty} f(x_1|x_2)dx_1 dx_2 = \int_0^2 \int_0^{x_1/2} k\, dx_2\, dx_1 = k\int_0^2 \frac{x_1}{2}dx_1 = \frac{k}{4}(x_1^2|_0^2)$

$= k$; i.e., $k = 1$.

b. $f(x_1) = \begin{cases} \int_0^{x_1/2} dx_2 = x_1/2 & 0 \le x_1 \le 2 \\ 0 & \text{otherwise} \end{cases}$

$f(x_2) = \begin{cases} \int_{2x_2}^2 dx_1 = 2(1-x_2) & 0 \le x_2 \le 1 \\ 0 & \text{otherwise} \end{cases}$

c. $f(x_1|x_2) = \dfrac{f(x_1, \quad x_2)}{f(x_2)} = \begin{cases} 1/2(1-x_2) & 0 \le 2x_2 \le x_1 \le 2 \\ 0 & \text{otherwise} \end{cases}$

d. $f(x_2|x_1) = \dfrac{f(x_1, \quad x_2)}{f(x_1)} = \begin{cases} \dfrac{1}{x_1/2} = 2/x_1 & 0 \le 2x_2 \le x_1 \le 2 \\ 0 & \text{otherwise} \end{cases}$

e. $P(X_1 \le 1.5, \quad X_2 < 0.5) = \int_0^{1/2} \int_{2x_2}^{3/2} dx_1 dx_2 = \int_0^{1/2} \left(\frac{3}{2} - 2x_2\right) dx_2$

$= \frac{3}{2}\left(\frac{1}{2}\right) - \frac{1}{4} = \frac{1}{2}$

f. $P\left(X_2 \le \frac{1}{2} \Big| X_1 \le \frac{3}{2}\right) = \dfrac{P\left(X_2 \le \frac{1}{2}, \quad X_1 \le \frac{3}{2}\right)}{P\left(X_1 \le \frac{3}{2}\right)} = \dfrac{\frac{1}{2}}{\int_0^{3/2} \frac{x_1}{2}dx_1} = \dfrac{\frac{1}{2}}{\frac{9}{16}} = \frac{8}{9}$

5.51

a. $E(X_1) = \int_0^1 x_1 2x_1 dx_1 = \dfrac{2}{3}$

b. $E(X_1^2) = \int_0^1 x_1^2 2x_1 dx_1 = \dfrac{1}{2}$

$V(X_1) = E(X_1^2) - [E(X_1)]^2 = \dfrac{1}{2} - \left(\dfrac{2}{3}\right)^2 = \dfrac{1}{18}$

c. $f(x_1, \quad x_2) = 4x_1 x_2 = f(x_1)f(x_2)$. Therefore, X_1 and X_2 are independent and $\text{Cov}(X_1, \quad X_2) = 0$.

5.53

a. $E(X_1 X_2) = 0\left(0 + \dfrac{1}{28} + \dfrac{1}{14} + \dfrac{1}{84} + \dfrac{1}{21} + \dfrac{1}{7} + \dfrac{1}{21}\right) + 1\left(\dfrac{2}{7}\right) + 2\left(\dfrac{3}{14} + \dfrac{1}{7}\right)$

$= \dfrac{2}{7} + \dfrac{6}{14} + \dfrac{2}{7} = 1$

$E(X_1) = 0\left(\dfrac{10}{84}\right) + 1\left(\dfrac{10}{21}\right) + 2\left(\dfrac{5}{14}\right) + 3\left(\dfrac{1}{21}\right) = 1.3333$

$E(X_2) = 0\left(\dfrac{5}{21}\right) + 1\left(\dfrac{15}{28}\right) + 2\left(\dfrac{3}{14}\right) + 3\left(\dfrac{1}{84}\right) = 1$

$\text{Cov}(X_1, \quad X_2) = E(X_1 X_2) - E(X_1)E(X_2) = 1 - 1.3333(1) = -0.3333$

b.

$x_1 + x_2$	$p(x_1 + x_2)$	
0	0	$= 0$
1	$\dfrac{1}{21} + \dfrac{1}{28}$	$= \dfrac{1}{12}$
2	$\dfrac{1}{14} + \dfrac{2}{7} + \dfrac{1}{7}$	$= \dfrac{1}{2}$
3	$\dfrac{1}{21} + \dfrac{1}{84} + \dfrac{1}{7} + \dfrac{3}{14}$	$= \dfrac{5}{12}$

$$E(X_1 + X_2) = 0(0) + 1\left(\frac{1}{12}\right) + 2\left(\frac{1}{2}\right) + 3\left(\frac{5}{12}\right) = \frac{7}{3}$$

$$E[(X_1 + X_2)^2] = 0^2(0) + 1^2\left(\frac{1}{12}\right) + 2^2\left(\frac{1}{2}\right) + 3^2\left(\frac{5}{12}\right) = \frac{35}{6}$$

$$V(X_1 + X_2) = E[(X_1 + X_2)^2] - [E(X_1 + X_2)]^2 = \frac{35}{6} - \left(\frac{7}{3}\right)^2 = \frac{7}{18}$$

c. $E(X_1 + X_2) = E(X_1) + E(X_2) = 1.3333 + 1 = 2.333$

$V(X_1 + X_2) = V(X_1) + V(X_2) + 2\text{Cov}(X_1, \quad X_2)$

Note that

$$E(X_1^2) = 1\left(\frac{10}{21}\right) + 4\left(\frac{5}{14}\right) + 9\left(\frac{1}{21}\right) = \frac{7}{3}$$

$$E(X_2^2) = 1\left(\frac{15}{28}\right) + 4\left(\frac{3}{14}\right) + 9\left(\frac{1}{84}\right) = \frac{3}{2}$$

so we have

$$V(X_1 + X_2) = E(X_1^2) - [E(X_1)]^2 + E(X_2^2) - [E(X_2)]^2 + 2\text{Cov}(X_1, \quad X_2)$$

$$= \frac{7}{3} - \left(\frac{4}{3}\right)^2 + \frac{3}{2} - 1^2 + 2\left(-\frac{1}{3}\right) = \frac{7}{18}.$$

5.55   Note that

$$E(X_1 X_2) = \int_0^2 \int_0^{x_1/2} x_1 x_2 dx_2 dx_1 = \int_0^2 \frac{x_1^3}{8} dx_1 = \frac{1}{32}(x_1^4|_0^2) = \frac{1}{2}$$

$$E(X_1) = \int_0^2 x_1 \left(\frac{x_1}{2}\right) dx_1 = \frac{4}{3}$$

$$E(X_1^2) = \int_0^2 x_1^2 \left(\frac{x_1}{2}\right) dx_1 = 2$$

$$E(X_2) = \int_0^1 x_2 2(1 - x_2) dx_2 = 2\left(\frac{1}{2} - \frac{1}{3}\right) = \frac{1}{3}$$

$$E(X_2^2) = \int_0^1 x_2^2 2(1 - x_2) dx_2 = 2\left(\frac{1}{3} - \frac{1}{4}\right) = \frac{1}{6}.$$

Therefore,

$$\text{Cov}(X_1, \quad X_2) = E(X_1X_2) - E(X_1)E(X_2) = \frac{1}{2} - \left(\frac{4}{3}\right)\left(\frac{1}{3}\right) = \frac{1}{18}$$

$$V(X_1) = E(X_1^2) - [E(X_1)]^2 = 2 - \left(\frac{4}{3}\right)^2 = \frac{2}{9}$$

$$V(X_2) = \frac{1}{6} - \left(\frac{1}{3}\right)^2 = \frac{1}{18}.$$

a. $E(X_1 + 2X_2) = E(X_1) + 2E(X_2) = \frac{4}{3} + 2\left(\frac{1}{3}\right) = 2$

b. $V(X_1 + 2X_2) = V(X_1) + 4V(X_2) + 2(2)\text{Cov}(X_1, \quad X_2) = \frac{2}{9} + 4\left(\frac{1}{18}\right) + 4\left(\frac{1}{18}\right)$

$$= \frac{2}{3}$$

5.57 $f(x|\lambda) = \dfrac{\lambda^x e^{-\lambda}}{x!}$

$$f(x, \lambda) = \left(\frac{\lambda^x e^{-\lambda}}{x!}\right) e^{-\lambda}$$

$$f(x) = \int_0^\infty \frac{\lambda^x e^{-2\lambda}}{x!} d\lambda = \frac{1}{x!} \int_0^\infty \lambda^x e^{-2\lambda} d\lambda$$

$$= \left(\frac{1}{x!}\right) \frac{\Gamma(x+1)}{2^{(x+1)}} \int_0^\infty \frac{1}{\Gamma(x+1)\left(\frac{1}{2}\right)^{(x+1)}} \lambda^x e^{-2\lambda} d\lambda$$

$$= \frac{\Gamma(x+1)}{x! 2^{(x+1)}} = \left(\frac{1}{2}\right)^{(x+1)} \quad \text{for } x = 0, 1, 2, \ldots$$

5.59 Let G = net daily gain = $X - Y$. Then

$$E(G) = E(X - Y) = E(X) - E(Y) = \mu - \alpha\beta = 50 - 4(2) = 42$$

$$V(G) = V(X) + V(Y) - 2\text{Cov}(X, \quad Y) = V(X) + V(Y) = \sigma^2 + \alpha\beta^2$$

$$= 10 + 4(2)^2 = 26$$

and, using Tchebysheff's theorem,

$$P(G > 70) = P[G - E(G) \geq 70 - E(G)] \leq P\left(|G - E(G)| \geq \frac{70 - E(G)}{\sqrt{V(G)}}\right.$$

$$\left. \times \sqrt{V(G)}\right) \leq \left(\frac{\sqrt{V(G)}}{70 - E(G)}\right)^2 = \frac{26}{(70 - 42)^2} = \frac{26}{784} = 0.03.$$

Therefore, it is unlikely that her net gain for tomorrow will exceed \$70.

5.61

  a. $E(X|\lambda) = \lambda$. Then

$$E(X) = E[E(X|\lambda)] = \int_{-\infty}^{\infty} E(X|\lambda)f(\lambda)d\lambda = \int_{2}^{\infty} \lambda e^{-\lambda}d\lambda = \Gamma(2) = 1.$$

  b. $E(X) = \sum_{x=0}^{\infty} xf(x) = \sum_{x=0}^{\infty} x\left(\frac{1}{2}\right)^{x+1} = \frac{1}{2}\sum_{x=0}^{\infty} x\left(\frac{1}{2}\right)^{x} = \frac{1}{2}\left\{\frac{\frac{1}{2}}{\left(\frac{1}{2}\right)^2}\right\} = 1$

5.63 $F(x_1) = \int_{0}^{x_1} \frac{1}{200}e^{-x/200}dx = 1 - e^{-x_1/200}, x_1 \geq 0$

  So, $f(x_2|x_1) = \dfrac{\frac{1}{200}e^{-x_2/200}}{1 - (1 - e^{-x_1/200})} = \frac{1}{200}e^{x_1 - x_2/200}, \quad x_2 \geq x_1$

  and $E(X_2|X_1 = 100) = \int_{100}^{\infty} x_2 \frac{1}{200}e^{100 - x_2/200}dx_2$

$$= \frac{e^{100/200}}{200}\int_{100}^{\infty} x_2 e^{-x_2/200}dx_2$$

$$= \frac{e^{100/200}}{200}\left(-200x_2 e^{-x_2/200} - 40{,}000e^{-x_2/200}\right)\Big|_{100}^{\infty}$$

$$= \frac{e^{100/200}}{200}\left(20{,}000e^{-100/200} + 40{,}000e^{-100/200}\right)$$

$$= 300.$$

5.65

a. $E\left(e^{t_1 X_1 + t_2 X_2 + t_3 X_3}\right)$

$$= \sum_{\substack{x_1, x_2, x_3 \\ x_1 + x_2 + x_3 = n}} e^{t_1 x_1 + t_2 x_2 + t_3 x_3} \frac{n!}{x_1! x_2! x_1!} p_1^{x_1} p_2^{x_2} p_1^{x_3}$$

$$= \sum_{\substack{x_1, x_2, x_3 \\ x_1 + x_2 + x_3 = n}} \frac{n!}{x_1! x_2! x_1!} \left(p_1 e^{t_1}\right)^{x_1} \left(p_2 e^{t_2}\right)^{x_2} \left(p_3 e^{t_3}\right)^{x_3}$$

$$= \left(p_1 e^{t_1} + p_2 e^{t_2} + p_3 e^{t_3}\right)^n$$

b. $E(X_1 X_2) = \left.\dfrac{\partial M_{X_1, X_2, X_3}(t_1, \quad t_2, \quad 0)}{\partial t_1 \partial t_2}\right|_{t_1 = t_2 = 0}$

$$= n p_1 (n-1) p_2 = n^2 p_1 p_2 - n p_1 p_2$$

$$E(X_1) = \left.\frac{\partial M_{X_1, X_2, X_3}(t_1, \quad 0, \quad 0)}{\partial t_1}\right|_{t_1 = 0} = n p_1$$

Similarly,

$E(X_2) = n p_2$

$\text{Cov}(X_1, \quad X_2) = E(X_1 X_2) - E(X_1) E(X_2) = n^2 p_1 p_2 - n p_1 p_2 - n p_1 n p_2$

$= -n p_1 p_2.$

5.67 $P(X_i = k) = \dfrac{1}{2}$ for $i = 1, 2, 3$, $k = 0, 1$, and

$$P(X_i = k, \quad X_{i'} = k') = \frac{1}{4} = P(X_i = k) P(X_{i'} = k')$$

for $i \neq i'$, $i = 1, 2, 3$; $i' = 1, 2, 3$; $k = 0, 1; k' = 0, 1$. Therefore, X_i and $X_{i'}$ are independent, for $i \neq i'$. However

$$P(X_1 = 1, \quad X_2 = 1, \quad X_3 = 1) = \frac{1}{4}$$

$$\neq P(X_1 = 1) P(X_2 = 1) P(X_3 = 1) = \frac{1}{8}.$$

Therefore, $X_1, \quad X_2, \quad X_3$ are not jointly independent.

Chapter 6.

6.1 a.

0		
·	76, 98	$Q_1 = (1.32 + 1.41)/2 = 1.365$
1	32, 41	$Q_2 = (1.77 + 1.84)/2 = 1.805$
·	56, 77, 84, 89, 90	$Q_3 = (1.90 + 2.01)/2 = 1.995$
2	01, 10, 42	*Note*: Where ·|76 denotes 0.76.

b. $\bar{x} = 19.96/12 = 1.6633$

$s^2 = (35.7092 - (19.96)^2/12)/11 = 0.2281$

c. Using Tchebysheff's theorem with $k = 2$, we estimate μ and σ by $\bar{x}$ and s, respectively, to get $(\bar{x} - 2s, \quad \bar{x} + 2s) = (0.7081, \quad 2.6185)$.

6.3

a. X = # displayed on die

X	# of occurrences in 30 tosses
1	5
2	5
3	5
4	5
5	5
6	5

b. $\bar{x} = 3.5 \quad s^2 = 3.0172$

6.5 Sample 1: $\overline{X}_1 = \dfrac{1}{5}(73.0 + 93.3 + 183.7 + 86.6 + 77.3) = 102.78$

$S_1^2 = \dfrac{1}{5 - 1} \left[(73.0)^2 + (93.3)^2 + (183.7)^2 + (86.6)^2 + (77.3)^2 - 5(102.78)^2 \right]$

$= 2108.95$

$$S_1 = \sqrt{S^2} = \sqrt{2108.95} = 45.92$$

$$\overline{X}_2 = 85.62 \qquad S_2 = 18.67$$

$$\overline{X}_3 = 89.44 \qquad S_3 = 10.97$$

$$\overline{X}_4 = 102.54 \qquad S_4 = 47.34$$

6.7 $\quad \displaystyle s^2 = \frac{1}{(n-1)} \sum_{i=1}^{n} (x_i - \bar{x})^2 = \frac{1}{(n-1)} \sum_{i=1}^{n} (x_i^2 - 2\bar{x}x_i - \bar{x}^2)$

$$= \frac{1}{(n-1)} \left\{ \sum_{i=1}^{n} x_i^2 - 2\bar{x} \sum_{i=1}^{n} x_i + n\bar{x}^2 \right\}$$

$$= \frac{1}{(n-1)} \left\{ \sum_{i=1}^{n} x_i^2 - 2\frac{\left(\sum_{i=1}^{n} x_i\right)^2}{n} + n\frac{\left(\sum_{i=1}^{n} x_i\right)^2}{n^2} \right\}$$

$$= \frac{1}{(n-1)} \left\{ \sum_{i=1}^{n} x_i^2 - \frac{1}{n} \left(\sum_{i=1}^{n} x_i\right)^2 \right\}$$

6.9 $\quad$ Let $\overline{X}$ = sample mean.

$$P(|\overline{X} - \mu| \le 1) = P(-1 \le \overline{X} - \mu \le 1) = P\left(\frac{-\sqrt{n}}{\sigma} \le \frac{\sqrt{n}(\overline{X} - \mu)}{\sigma} \le \frac{\sqrt{n}}{\sigma} \right)$$

$$\approx P\left(-\frac{\sqrt{100}}{10} \le Z \le \frac{\sqrt{100}}{10} \right) = P(-1 \le Z \le 1) = 2(0.3413) = 0.6826.$$

6.11 $\quad$ Let $\bar{X}$ = sample mean. Since, for mound-shaped symmetric distributions, 95% of the observations lie within 2 standard deviations of the mean, assume the standard deviation of the pH to be one fourth of the usual range; i.e., $\sigma = (8-5)/4 = 3/4$. Then

$$P(|\bar{X} - \mu| \le 0.2) = P\left(-\frac{0.2\sqrt{40}}{(3/4)} \le \frac{(\bar{X} - \mu)\sqrt{40}}{(3/4)} \le \frac{0.2\sqrt{40}}{(3/4)} \right)$$

$$\simeq P(-1.687 \le Z \le 1.687) = 2(0.4545) = 0.9090.$$

6.13

a. $P(199 < \bar{X} < 202) = P\left(\dfrac{\sqrt{25}(199 - 200)}{10} < \dfrac{\sqrt{n}(\bar{X} - \mu)}{\sigma} < \dfrac{\sqrt{25}(202 - 200)}{10} \right)$

$\approx P(-0.5 < Z < 1) = 0.1915 + 0.3413 = 0.5328$

b. $P\left(\sum\limits_{i=1}^{25} X_i \leq 5100 \right) = P\left(\dfrac{\sum\limits_{i=1}^{25} X_i}{25} \leq \dfrac{5100}{25} \right) = P(\bar{X} \leq 204)$

$= P\left(\dfrac{\sqrt{n}(\bar{X} - \mu)}{\sigma} \leq \dfrac{\sqrt{25}(204 - 200)}{10} \right) \approx P(Z \leq 2) = 0.5 + 0.4772 = 0.9772$

c. The approximations in parts (a) and (b) assume that the resistances are independent, and random sampling.

6.15 $P(\bar{X} < 1.3) = P\left(\dfrac{\sqrt{n}(\bar{X} - \mu)}{\sigma} < \dfrac{\sqrt{25}(1.3 - 1.4)}{0.5} \right) \approx P(Z < -1) \approx 0.1587$

6.17

a. $P\left(\sum\limits_{i=1}^{100} X_i > 45 \right) = P(\bar{X} > 0.45) \approx P\left(Z > \dfrac{\sqrt{100}(0.45 - 0.5)}{0.2} \right)$

$= P(Z > -2.5)$

$= 0.5 + 0.4938 = 0.9938$

b. $P\left(\sum\limits_{i=1}^{100} X_i > 50 \right) = P(\bar{X} > \dfrac{50}{n}) \approx P\left(Z > \dfrac{\sqrt{n}((50/n) - 0.5)}{0.2} \right) = 0.99$

This equation is true for $\dfrac{\sqrt{n}\,((50/n) - 0.5)}{0.2} = -2.33$; i.e., $0.5n - 0.466\sqrt{n} -$

$50 = 0$. Using the quadratic formula, we have $\sqrt{n} = \dfrac{0.466 \pm \sqrt{(0.466)^2 - 4(0.5)(-50)}}{2(0.5)}$

$= -9.5449$ or 10.4769. Using the positive root, we have $n = (10.4769)^2$

$= 109.77$; i.e., $n = 110$.

6.19   Let $X_i =$ service time for i^{th} customer, $i = 1, \ldots, 100$.

$$P\left(\sum_{i=1}^{100} X_i < 120\right) = P\left(\bar{X} < \frac{120}{100}\right) = P\left(\bar{X} < 1.2\right)$$

$$= P\left(\frac{\sqrt{n}(\bar{X} - \mu)}{\sigma} < \frac{\sqrt{100}(1.2 - 1.5)}{1}\right)$$

$$\approx P(Z < -3) = 0.5 - 0.4987 = 0.0013$$

6.21 $P\left(\left|\bar{X} - \bar{Y} - (\mu_1 - \mu_2)\right| \le 0.05\right) =$

$$P\left\{\left|\bar{X} - \bar{Y} - (\mu_1 - \mu_2)/\sqrt{(\sigma_1^2/n_1) + (\sigma_2^2/n_2)}\right| \le \frac{0.05}{\sqrt{(0.01/50) + (0.02/100)}}\right\}$$

$$\approx P(|Z| \le 2.5) = 2P(0 \le Z \le 2.5) = 2(0.4938) = 0.9876$$

6.23   Let $\bar{X}$ and $\bar{Y}$ be the sample means for A and B, respectively. Then

$$P(\bar{Y} - \bar{X} > 1) = P\left\{\bar{Y} - \bar{X} - 0/\sqrt{(\sigma^2/n) + (\sigma^2/n)} > \frac{1-0}{\sqrt{2(4/50)}}\right\}$$

$$\approx P(Z > 2.5) = 0.5 - 0.4938 = 0.0062.$$

6.25   Let Y = number of nonconformances out of the sample of 50. Then Y has a binomial distribution with parameters $n = 50$, p. Let X be a normally distributed random variable with parameters $\mu = np = 50p$, $\sigma^2 = np(1 - p) = 50p(1 - p)$. Then

$$\text{P(accepting the lot)} = P(Y \le 5) \approx P(X \le 5.5) = P\left(\frac{X - \mu}{\sigma} \le \frac{5.5 - 50p}{\sqrt{50p(1-p)}}\right)$$

$$= P\left(Z \le \frac{5.5 - 50p}{\sqrt{50p(1-p)}}\right).$$

a. For $p = 0.1$, we have

$$P(Y \le 5) \approx P\left(Z \le \frac{5.5 - 50(0.1)}{\sqrt{50(0.1)(1 - 0.1)}}\right) = P(Z \le 0.24)$$

$$= 0.5948.$$

b. For $p = 0.2$, we have

$$P(Y \leq 5) \approx P\left(Z \leq \frac{5.5 - 50(0.2)}{\sqrt{50(0.2)(1 - 0.2)}}\right) = P(Z \leq -1.59) = 0.5 - 0.4441$$

$$= 0.0559.$$

c. For $p = 0.3$, we have

$$P(Y \leq 5) \approx P\left(Z \leq \frac{5.5 - 50(0.3)}{\sqrt{50(0.3)(1 - 0.3)}}\right) = P(Z \leq -2.93) = 0.5 - 0.4983$$

$$= 0.0017.$$

6.27   Let Y = number of disks that contain missing pulses out of the sample of 100. Then Y has a binomial distribution with parameters $n = 100$, $p = 0.2$. Let X be a normally distributed random variable with parameters $\mu = np = 100(0.2) = 20, \sigma^2 = np(1 - p) = 100(0.2)(0.8) = 16$. Then

$$P(Y \leq 15) \approx P(X \leq 15.5) = P\left(\frac{X - \mu}{\sigma} \leq \frac{15.5 - 20}{\sqrt{16}}\right)$$

$$= P(Z \leq -1.125) = 0.5 - 0.3708 = 0.1292.$$

6.29   Let Y = number of days in which demand for over 500,000 gallons per day occurred, out of the sample of 30. Then Y has a binomial distribution with parameters $n = 30$, $p = 0.15$. Let X be a normally distributed random variable with parameters $\mu = np = 30(0.15) = 4.5$, $\sigma^2 = np(1 - p) = 30(0.15)(0.85) = 3.825$. Then

$$P(Y \leq 2) \approx P(X \leq 2.5) = P\left(\frac{X - \mu}{\sigma} \leq \frac{2.5 - 4.5}{\sqrt{3.825}}\right)$$

$$= P(Z \leq -1.02) = 0.5 - 0.3461 = 0.1539.$$

6.31   Let W = waiting time. Then

$$P(W > 10) = 1 - F(10) = 1 - \left(1 - e^{-10/10}\right) = e^{-1}$$

Let Y = number of customers whose waiting time exceeds 10 minutes, out of the sample of 100. Then Y has a binomial distribution with parameters $n = 100$, $p = P(W > 10) = e^{-1}$. Let X be a normally distributed random variable with parameters $\mu = np = 100e^{-1} = 36.7880$, $\sigma^2 = np(1 - p) =$

$100e^{-1}(1 - e^{-1}) = 23.2544$. Then

$$P(Y \geq 50) \approx P(X \geq 49.5) = P\left(\frac{X - \mu}{\sigma} \geq \frac{49.5 - 36.7880}{\sqrt{23.2544}}\right)$$

$$= P(Z \geq 2.6361)$$

$$= 0.5 - 0.4959 = 0.0041.$$

6.33 Let Y = number of vouchers that show up as being improperly documented out of the sample of 100. Then Y has a binomial distribution with parameters $n = 100$, $p = 0.2$. Let X be a normally distributed random variable with parameters $\mu = np = 100(0.2) = 20$, $\sigma^2 = np(1 - p) = 100(0.2)(0.8) = 16$.

$$P(Y > 30) \approx P(X > 30.5) = P\left(\frac{X - \mu}{\sigma} > \frac{30.5 - 20}{\sqrt{16}}\right) = P(Z > 2.63)$$

$$= 0.5 - 0.4957 = 0.0043$$

6.35 $P(\bar{x} > 10) = P\left(Z > \dfrac{10 - \mu}{0.5/\sqrt{8}}\right) = P(Z > z_0) = 0.90$. Hence, $z_0 = \dfrac{10 - \mu}{0.5/\sqrt{8}} = 1.28$

so that $\mu = 10 + \dfrac{1.28(0.5)}{\sqrt{8}} = 10.23$.

6.37 $P(|\bar{x} - \mu| \leq 0.5) = P\left(|Z| < \dfrac{0.5}{\sqrt{1.9}/\sqrt{n}}\right) = P(|Z| \leq z_0) = 0.95$.

Thus $z_0 = \dfrac{0.5}{\sqrt{1.9}/\sqrt{n}} = 1.96$ so that $n = 29.20$. Rounding to obtain an approximate 0.9 probability, we choose $n = 29$.

6.39 Begin by assuming that the specification variance of $(0.2)^2 = 0.04$ is the true value of σ^2. If the test circuit variance of 0.065 is so far from 0.04 that we would not expect it to be this large by chance, then we have evidence against our assumption that $\sigma^2 = 0.04$. However, if 0.065 is close enough to 0.04 so that it would be likely to be observed by chance, then we have no reason to doubt the assumption that $\sigma^2 = 0.04$. A probability less than, say, 0.05 is usually taken to indicate that the sample value "significantly" differs from the specified value so much that it gives evidence against the assumed specified value. In our case, we assume $\sigma^2 = 0.04$ and we examine the probability of obtaining a sample value of 0.065 or even larger (since larger values would give even more evidence against $\sigma^2 = 0.04$). Assuming a normal population, we have

$$P(s^2 \geq 0.065) = P\left(\frac{s^2(n-1)}{\sigma^2} \geq \frac{0.065(10-1)}{0.04}\right) = P(U \geq 14.625).$$

Note that, with 9 degrees of freedom, we find from the χ^2 table that $P(U > 14.6837) = 0.1$. Thus, $P(s^2 \geq 0.065) \approx 0.1$. Hence, we would expect to get a sample value of 0.065 or larger about 10% of the time by chance if $\sigma^2 = 0.04$. This is not rare enough (less than 0.05 probability) for us to conclude that σ^2 is greater than 0.04. We will not reject the specification of $\sigma^2 = 0.04$.

6.41

    a. $P(S^2 > 80) = P\left(U > \frac{(80)(24)}{50}\right) = P(U > 38.4)$. From χ^2 table, with 24 degrees of freedom, we find $0.025 < P(S^2 > 80) < 0.050$.

    b. $P(S^2 < 20) = P\left(U < \frac{(20)(24)}{50}\right) = P(U < 9.6) < 0.005$

    c. $E(S^2) = \sigma^2 = 50, \quad V(S^2) = \frac{2\sigma^4}{(n-1)} = \frac{2(50)^2}{(24)} = 208.3333$. Tchebysheff's theorem then gives the interval as $\left(50 - 2\sqrt{208.3333}, \quad 50 + 2\sqrt{208.3333}\right) =$ (21.13,   78.87).

    d. Assume the population of resistances is approximately normal.

6.43   Let a and b be the endpoints of the desired interval. Then $P\left(\frac{19a}{225} \leq \chi^2 \frac{19b}{225}\right)$

$= 0.9$. Consequently, $\frac{19a}{225} = 10.1170$ so that $a = 119.8066$; $\frac{19b}{225} = 30.1435$ giving $b = 356.9625$.

6.45

    a. $\bar{\bar{x}} = 15.0283, \quad \bar{R} = 0.3467, \quad A_2 = 0.729$

    Control limits: $15.0283 \pm 0.729(0.3467) = (14.776, 15.281)$

b. Since samples 9, 10, and 12 have means outside the Control limits, we recalculate $\bar{\bar{x}} = 15.0104$ and $\bar{R} = 0.3417$.

Control limits: $15.0104 \pm 0.729(0.3417) = (14.761, 15.260)$

6.47 Control limits: $\bar{\bar{x}} \pm \dfrac{3\bar{s}'}{c_2} = 15.0283 \pm \dfrac{3(0.1362)}{0.7979} = (14.5162, 15.5404)$ where c_2 is found from Table 12 in the Appendix. Since this interval is wider than $15 \pm 0.4 = (14.6, 15.4)$, the individual readings do not seem to be meeting the specifications.

6.49

a. $\bar{P} = \dfrac{137}{100(20)} = 0.0685$

Control limits: $0.0685 \pm 3\sqrt{\dfrac{(0.0685)(0.9315)}{100}}$ or $(0, 0.1443)$

b. Since $p_{11} = 15/100 = 0.15$ does not fall within the control limits, sample 11 is omitted and the new control limits for $\bar{P} = \dfrac{122}{100(19)} = 0.0642$ are given by

$0.0642 \pm 3\sqrt{(0.0642)(0.9358)/100}$ or $(0, 0.1377)$. Since $p_4 = 0.14$ is outside these control limits, we omit sample 4 and recompute $\bar{P} = \dfrac{108}{100(18)} = 0.06$.

The control limits are now $0.06 \pm 3\sqrt{\dfrac{(0.06)(0.94)}{100}}$ or $(0, 0.1312)$, and all remaining samples have sample proportions within this interval.

6.51

a. $\bar{P} = \dfrac{116}{50(30)} = 0.0773$

Control limits: $0.0773 \pm 3\sqrt{\dfrac{(0.0773)(0.9227)}{50}}$ or $(0, 0.1906)$

b. Since $p_7 = 10/50 = 0.2$ is outside the control limits, we delete sample 7 and recompute $\bar{P} = \dfrac{106}{50(29)} = 0.0731$ to get the new control limits

$0.0731 \pm 3\sqrt{\dfrac{(0.0731)(0.9269)}{50}}$ or $(0, 0.1835)$.

6.53 Control limits: $5.67 \pm 2.575\sqrt{5.67}$ or $(0, 11.8015)$

6.55 Since λ is both the mean and variance for the Poisson (λ) distribution, control limits for λ may be simultaneously viewed as limits for the mean or variance. Hence, a separate method is not necessary.

6.57 C_{p_k} will increase. Proportion will decrease.

6.59 Exercise for student.

6.61 Note that $a = \dfrac{2d}{t^2}$. To approximate the average acceleration $\bar{a}$, we consider $g(\bar{t}) = \dfrac{2d}{(\bar{t})^2}$. Approximations for the mean and variance are now given by the following:

Mean of $\bar{a}$: $g(\mu) = \dfrac{2d}{\mu^2} = \dfrac{2(10)}{(20)^2} = 0.05$ where μ is the true mean time

Variance of $\bar{a}$: $\dfrac{g'(\mu)^2 \sigma^2}{n} = \dfrac{\left(\mu^{-4d/3}\right)^2 \left(\sigma^2\right)}{n} = \left(\dfrac{-4(10)}{(20)^3}\right)^2 \left(\dfrac{(1.6)^2}{100}\right)$

$= 0.0000006$

6.63 Since population ages and incomes are likely to be skewed, medians are appropriate. On the other hand, SAT scores are likely to be symmetric — in fact, are probably normal, — so that the mean score is appropriate.

6.65 Note that each X_i is distributed beta $(\alpha = 3, \quad \beta = 1)$ so that $E(X_i) = \dfrac{\alpha}{(\alpha + \beta)}$

$= \dfrac{3}{(3+1)} = 0.75$ and $V(X_i) = \dfrac{\alpha\beta}{((\alpha+\beta)^2(\alpha+\beta+1))} = \dfrac{3(1)}{((3+1)^2(3+1+1))}$

$= 0.0375$. Hence,

$P(\bar{x} > 0.7) = P\left(Z > \dfrac{0.7 - 0.75}{\dfrac{\sqrt{0.0375}}{\sqrt{40}}} \right) = P(Z > -1.63) = 0.5 + 0.4483$

$= 0.9484.$

6.67 Let $W = \bar{x}_a - \bar{x}_b$ so that W is normal $(\mu = \mu_a - \mu_b = 0, \quad \sigma^2 = \dfrac{\sigma_a^2}{n_a} + \dfrac{\sigma_2^2}{n_b}$

$$= \frac{1.9}{10} + \frac{0.8}{10} = 0.27).$$

$$P(\bar{x}_a - \bar{x}_b \geq 1) = P(W \geq 1) = P\left(Z \geq \frac{1-0}{\sqrt{0.27}}\right) = P(Z \geq 1.92)$$

$$= 0.5 - 0.4726 = 0.0274$$

6.69 Since each of the Y_i, i=1, ..., 5, is exponential $(\theta = 20)$, we use Exercise 6.68

to conclude that $\dfrac{2Y_i}{20} = \dfrac{Y_i}{10}, \; i = 1, \ldots, 5$, is distributed chi-square $(\nu = 2)$

or, equivalently, gamma $(\alpha = 1, \quad \beta = 2)$. Since the sum of gamma random

variables still has a gamma distribution, we have $\displaystyle\sum_{i=1}^{5} \frac{Y_i}{10}$ is distributed gamma $(\alpha =$

$\dfrac{10}{2} = 5, \quad \beta = 2)$ or, equivalently, chi-square $(\nu = 10)$. Hence, $P\left(\displaystyle\sum_{i=1}^{5} Y_i > c\right) =$

$P\left(\displaystyle\sum_{i=1}^{5} Y_i/10 > c/10\right) = P\left(U > \dfrac{c}{10}\right) = 0.05$. By χ^2 table, we find $\dfrac{c}{10} = \chi^2_{0.05} =$

18.3070 so that c = 183.070.

Chapter 7.

7.1

a. $E(\hat{\theta}_1) = E(X_1) = \theta$

$$E(\hat{\theta}_2) = E\left(\frac{X_1 + X_2}{2}\right) = \left(\frac{E(X_1) + E(X_2)}{2}\right) = \frac{\theta + \theta}{2} = \theta$$

$$E(\hat{\theta}_3) = E\left(\frac{X_1 + 2X_2}{3}\right) = \left(\frac{E(X_1) + 2E(X_2)}{3}\right) = \frac{\theta + 2\theta}{3} = \theta$$

$E(\hat{\theta}_4) = E(\bar{x}) = \theta$

Hence, each of the four estimators is unbiased for θ.

b. $V(\hat{\theta}_1) = V(X_1) = \theta^2$

$$V(\hat{\theta}_2) = V\left(\frac{X_1 + X_2}{2}\right) = \left(\frac{V(X_1) + V(X_2)}{4}\right) = \frac{\theta^2 + \theta^2}{4} = \frac{\theta^2}{2}$$

$$V(\hat{\theta}_3) = V\left(\frac{X_1 + X_2}{3}\right) = \left(\frac{V(X_1) + 4V(X_2)}{9}\right) = \frac{\theta^2 + 4\theta^2}{9} = \frac{5\theta^2}{9}$$

$$V(\hat{\theta}_4) = V(\bar{x}) = \frac{\sigma^2}{n} = \frac{\theta^2}{3}$$

Thus, $\hat{\theta}_4 = \bar{x}$ has the smallest variance among these four estimators.

7.3

a. Since λ is the mean of a Poisson (λ) distribution, $\bar{x}$ will be an unbiased estimator of λ; i.e., $E(\bar{x}) = \mu = \lambda$.

b. $E(C) = E(EY + Y^2) = 3E(Y) + E(Y^2) = 3\lambda + (V(Y) + (E(Y))^2)$

$= 3\lambda + (\lambda + \lambda^2) = \lambda^2 + 4\lambda$

c. Consider $\frac{1}{n}\sum_{i=1}^{n} X_i^2 + 3\bar{x}$. Since $E\left(\frac{1}{n}\sum_{i=1}^{n} X_i^2 + 3\bar{x}\right) = \frac{1}{n}\sum_{i=1}^{n} E(X_i^2) + 3\bar{x}$

$= \frac{1}{n}\sum_{i=1}^{n}(V(X_i) + (E(X_i))^2) + 3\lambda = \frac{1}{n}\sum_{i=1}^{n}(\lambda + \lambda^2) + 3\lambda = \lambda^2 + 4\lambda = E(C)$.

This estimator is unbiased for E(C).

7.5 $\mathrm{MSE}(\bar{x}) = V(\bar{x}) + B^2 = \sigma^2/n + (\theta - E(\bar{x}))^2 = \dfrac{(\theta + 1 - \theta)^2}{12n} + \left(\theta - \left(\theta + \dfrac{1}{2}\right)\right)^2$

$$= \dfrac{1}{12n} + \dfrac{1}{4}$$

7.7 Since $\bar{x}$ is an unbiased estimator of μ and Exercise 7.6 provides an unbiased estimator of σ, we have $\bar{x} - 1.645s\sqrt{(n-1)/2}\,\Gamma((n-1)/2)/\Gamma(n/2)$ is an unbiased estimator of $\mu - 1.645\sigma$.

7.9 With $\bar{x} = 210$, $\quad s = 18$, and $n = 50$, we compute a large sample 90% confidence interval for the mean breaking strength to be

$$\bar{x} \pm z_{0.05}\sigma/\sqrt{n} \approx 210 \pm 1.645(18)/\sqrt{50} = (205.81, \quad 214.19).$$

7.11 We seek a 98% large sample confidence interval for the mean percent of shrinkage given that $\bar{x} = 18.4$, $\quad s = 1.2$, and $n = 45$. Thus,

$$x + z_{0.01}\sigma/\sqrt{n} \approx 18.4 \pm 2.33(1.2)/\sqrt{45} = (17.98, \quad 18.82).$$

7.13 The sample size for estimating the mean hours worked to within $B = 0.5$ hours with confidence coefficient $1 - \alpha = 0.95$ and $\sigma \approx 3$ hours is given by

$$n = (z_{0.05}\sigma/B)^2 \approx (1.96(3)/0.5)^2 = 138.3 \text{ or } n = 139.$$

7.15 With $\hat{p} = y/n = 12/100$, an approximate 95% confidence interval for the population proportion of resistors that fail to meet the tolerance specification is as follows:

$$\hat{p} \pm z_{0.0252}\sqrt{\hat{p}(1 - \hat{p})/n} = 0.12 \pm 1.96\sqrt{(0.12)(0.88)/100} = (0.0563, \quad 0.1837).$$

7.17 An approximate 98% confidence interval for the true proportion of supports with hairline cracks where $\hat{p} = 28/70 = 0.4$ is

$$\hat{p} \pm z_{0.01}\sqrt{\hat{p}(1 - \hat{p})/n} = 0.4 \pm 2.33\sqrt{(0.4)(0.6)/70} = (0.2636, \quad 0.5364).$$

7.19 An approximate 90% confidence interval for the proportion of stocked items for which the audit value exceeds the book value is as follows for $\hat{p} = 45/60 = 0.75$:

$$p \pm z_{0.05}\sqrt{\hat{p}(1 - \hat{p})/n} = 0.75 \pm 1.645\sqrt{(0.75)(0.25)/60} = (0.6580, \quad 0.8420).$$

7.21　We are given that $\bar{x} = 180$,　$s = 5$, and $n = 5$. Assuming a normal population of warpwise breaking-strength measurements, a 95% confidence interval for the true mean warpwise breaking-strength is

$$\bar{x} \pm \frac{t_{0.05}s}{\sqrt{n}} = 180 \pm \frac{2.776(5)}{\sqrt{5}} = (173.7927,　186.2073) \text{ where } t_{0.025} \text{ is found}$$

from t-table, with 4 degrees of freedom.

7.23　Here $\bar{x} = 9.8$,　$s = 0.5$, and $n = 15$. A 95% confidence interval for the true mean resistance is given by $\bar{x} \pm t_{0.05}s/\sqrt{n} = 9.8 \pm 2.145(0.5)/\sqrt{15} = (9.5231,　10.0769)$ where $t_{0.025}$ is found from t-table, with 14 degrees of freedom.

7.25　We are given $s^2 = (0.5)^2 = 0.25$ and $n = 15$. Using χ^2 table, with 14 degrees of freedom, and assuming a normal population of resistance measurements, a 90% confidence interval for the variance of the resistances is

$$\left(\frac{(n-1)s^2}{x_{0.05}^2(14)},　\frac{(n-1)s^2}{x_{0.95}^2(14)} \right) = \left(\frac{14(0.25)}{23.6848},　\frac{14(0.25)}{6.57063} \right)$$
$$= (0.1478,　0.5327).$$

7.27　We are told that $\bar{x} = 477$,　$s = 13$, and $n = 150$. Then a 90% confidence interval for the true mean yield stress is found to be

$$\bar{x} \pm x_{0.05}\sigma/\sqrt{n} \approx 477 \pm 1.645(13)/\sqrt{150} = (475.2539,　478.7461).$$

No assumptions are necessary.

7.29　From the data, we compute $\sum_{i=1}^{9} x_i = 4,950$, $\sum_{i=1}^{9} x_1^2 = 3,195,442$ and $n = 9$ so that

$\bar{x} = 4,950/9 = 550$ and $s = \left((3,195,442 - (4,950)^2/9)8 \right)^{1/2} = 243.1414$. A 90% confidence interval for the mean number of cycles to failure is

$$\bar{x} \pm t_{0.05}s/\sqrt{n} = 550 \pm 1.860(243.1414)/\sqrt{9} = (399.2523,　700.7477)$$

where $t_{0.05}$ with 8 degrees of freedom is found from t-table.

7.31   The 3% was calculated as the sampling error (or half the width of the confidence interval). Hence, $0.03 = z_{\alpha/2}\sqrt{(0.51)(0.49)/1,000}$ so that $z \approx 1.90$. The confidence coefficient $1 - \alpha = P(-190 \leq z \leq 1.90) = 2(0.4731) = 0.9426$. This corresponds to a 94.26% confidence level. Due to probable rounding in the figure 0.03 as the sampling error, the confidence level should most likely be rounded to 95%. Thus, the sampling error of 3% can be interpreted as giving half the width of a 95% confidence interval for the true proportion of registered voters who agree with the statement.

Since the confidence interval $(\hat{p}$ ± sampling error$) = (0.51 \pm 0.03) = (0.48, 0.54)$ includes values less than 0.5, we cannot conclude that a majority of registered voters agree with the statement.

7.33   For a sample of size $n = 1,200$, the sampling error for estimating a proportion is $z_{\alpha/2}\sqrt{p(1-p)}$. Note that this sampling error depends on the sample size n, but *not* on the size of the population. Hence, even if the population size is 80 million rather than 100,000, the sampling error remains unchanged for samples of size n. Since the smaller population size of 100,000 is still large enough to yield a good estimate of p, even the estimated standard error (using $\hat{p}$ in place of p) will be relatively unchanged.

7.35   We need (half the width of the confidence interval) $\leq$ bound or $z_{0.025}(\sigma_1^2/n_1 + \sigma_2^2/n_2) \leq B$. Substituting, we have $1.96\left((20 + 30)/n\right)^{\frac{1}{2}} \leq 1$. Solving for n gives $n \geq (1.96)^2(20 + 30) = 192.08$ so that $n = 193$.

7.37   We are given $\hat{p}_1 = 43/50 = 0.86$ and $\hat{p}_2 = 22/50 = 0.44$. An approximate 95% confidence interval for the true difference between the proportions of samples containing the harmful bacteria is

$$\hat{p}_1 - \hat{p}_2 \pm z_{0.025}\sqrt{\frac{\hat{p}_1(1 - \hat{p}_1)}{n_1} + \frac{\hat{p}_2(1 - \hat{p}_2)}{n_2}}$$

$$= 0.86 - 0.44 \pm 1.96 \sqrt{\frac{(0.86)(0.14)}{50} + \frac{(0.44)(0.56)}{50}}$$

$$= (0.2521, \quad 0.5879).$$

If the additive is effective in reducing the amount of bacteria, then the difference between the proportions should be positive; i.e. $p_1 - p_2 > 0$. The above 95% confidence level is positive indicating that the chemical is effective in reducing the amount of bacteria.

7.39   A 98% confidence interval for the difference between the proportions voicing no objection to the new policy for the two companies is

$$\hat{p}_1 - \hat{p}_2 \pm z_{0.01} \sqrt{\frac{p_1(1 - p_1)}{n_1} + \frac{p_2(1 - p_2)}{n_2}} \approx \frac{25}{30} - \frac{25}{35}$$

$$\pm \quad 2.33 \sqrt{\frac{(25/30)(5/30)}{30} + \frac{(25/35)(10/35)}{35}} = (-0.1193, \quad 0.3574).$$

7.41   From the given data, we have

	DDT	Diazinon
n	12	3
$\bar{x}$	9	3.5667
s	6.4244	3.6692

$$s_p^2 = \frac{11(6.4244)^2 + 2(3.6692)^2}{12 + 3 - 2} = 36.9945 \qquad s_p = 6.08230$$

Assuming normal populations with equal variances, then a 90% confidence interval for the difference between the mean LC50 for DDT and Diazinon with degrees of freedom = 13 is

$$\bar{x}_1 - \bar{x}_2 \pm t_{0.05} s_p \sqrt{\frac{1}{n_1} + \frac{1}{n_2}} = 9 - 3.5667 \pm 1.771(6.08231) \sqrt{\frac{1}{12} + \frac{1}{3}}$$

$$= (-1.5198, \quad 12.3865).$$

A 90% confidence interval for the true variance ratio, using F table, is

$$\left(\frac{s_1^2}{s_2^2} \frac{1}{F_{0.05}(11, 2)}, \quad \frac{s_1^2}{s_2^2} F_{0.05}(2, \quad 11) \right) = \left(\frac{(6.4244)^2}{(3.6692)^2} \frac{1}{19.405}, \quad \frac{(6.4244)^2}{(3.6692)^2}(3.98) \right)$$

$$= (0.1580, \quad 12.2013).$$

7.43 From the data, we compute the following summary statistics:

	Spring	Summer
n	6	4
$\bar{x}$	16.0333	72.275
s	8.9146	24.1300

$$s_p^2 = \frac{5(8.9146)^2 + 3(24.1300)^2}{6 + 4 - 2} = 268.01515 \qquad s_p = 16.3712.$$

With degrees of freedom = 8, a 95% confidence interval for the difference between mean spring and summer ranges is

$$\bar{x}_1 - \bar{x}_2 \pm t_{0.025}s_p\sqrt{\frac{1}{n_1} + \frac{1}{n_2}} = 16.0333 - 72.275 \pm 2.306(16.3712)\sqrt{\frac{1}{6} + \frac{1}{4}}$$

$$= (-80.6105, \quad -31.8729).$$

7.45 We compute $s_p^2 = \dfrac{(n_1 - 1)s_1^2 + (n_2 - 1)s_2^2}{n_1 + n_2 - 2} = \dfrac{44(82.4) + 18(38.1)}{45 + 19 - 2} = 69.5387$ so

that $s_p = 8.3390$. Assuming normal population and equal variances, we have

$$\bar{x}_1 - \bar{x}_2 \pm t_{0.025}s_p\sqrt{\frac{1}{n_1} + \frac{1}{n_2}} \qquad \text{(degrees of freedom = 62)}$$

$$= 11.6 - 6.2 \pm 1.96(8.3390)\sqrt{\frac{1}{45} + \frac{1}{19}} = (0.9283, \quad 9.8717).$$

7.47 From the data, we calculate

	Method A	Method B
n	7	8
$\bar{x}$	0.6714	1.15
s	0.2059	0.6990

$$s_p^2 = \frac{6(0.2059)^2 + 7(0.6990)^2}{7 + 8 - 2} = 0.2827 \qquad s_p = 0.5317$$

Assuming that both populations of corrosion rates are approximately normal with common variance, a 90% confidence interval for the difference in mean corrosion rates is

$$\bar{x}_1 - \bar{x}_2 + t_{0.05}s_p\sqrt{\frac{1}{n_1} + \frac{1}{n_2}} = 0.6714 - 1.15 \pm 1.771(0.5317)\sqrt{\frac{1}{7} + \frac{1}{8}}$$

$$= (-0.9959, 0.0087).$$

7.49   From the given data, we have

	Seawater	Air
n	9	9
$\bar{x}$	550	583.2222
s	243.1414	175.1205

$$s_p^2 = \frac{8(243.1414)^2 + 8(175.1205)^2}{9 + 9 - 2} = 44{,}892.464 \qquad s_p = 211.8784.$$

Assuming normal populations and equal variances, a 95% confidence interval for the difference between mean cycles to failure for seawater and air is

$$\bar{x}_1 - \bar{x}_2 \pm t_{0.025}s_p\sqrt{\frac{1}{n_1} + \frac{1}{n_p}} = 550 - 583.2222 \pm 2.120(211.8784)\sqrt{\frac{1}{9} + \frac{1}{9}} =$$

$$(-244.9688, 178.5243).$$

If seawater lessens the mean number of cycles to failure, then $\mu_1 - \mu_2 < 0$. Since the confidence interval above includes 0, we cannot conclude that seawater lessens the mean number of cycles to failure.

7.51   Using $\hat{p} = \frac{32}{400} = 0.08$ and $\hat{p}_2 = \frac{28}{400} = 0.07$, an approximate 95% confidence interval for the difference between proportions of improperly broken microchips for the two methods is

$$\hat{p}_1 - \hat{p}_2 \pm z_{0.025}\sqrt{\frac{\hat{p}_1(1 - \hat{p}_1)}{n_1} + \frac{\hat{p}_2(1 - \hat{p}_2)}{n_2}}$$

$$= 0.08 - 0.07 \pm 1.96\sqrt{\frac{(0.08)(0.92)}{400} + \frac{(0.07)(0.93)}{400}} = (-0.0265, 0.0465).$$

Since this confidence interval includes the value 0 for an estimate of $p_1 - p_2$, we cannot conclude that either method is superior using this 95% confidence interval.

7.53 A point estimate of $\mu_I + \mu_{II} + \mu_{III}$ is given by $\bar{x}_1 + \bar{x}_2 + \bar{x}_3 = 9.1 + 14.3 + 5.6 = 29$.

Assuming a common population variance for the three populations of resistances, we compute

$$s_p^2 = \frac{(n_1 - 1)s_1^2 + (n_2 - 1)s_2^2 + (n_3 - 1)s_3^2}{n_1 + n_2 + n_3 - 3} = \frac{9(0.2)^2 + 7(0.4)^2 + 11(0.1)^2}{10 + 8 + 12 - 3}$$

$$= 0.05889 \qquad s_p = 0.2427.$$

Then, assuming all three resistance populations are normal, a 95% confidence interval for $\mu_I + \mu_{II} + \mu_{III}$ with 27 degrees of freedom is

$$\hat{\theta} \pm t_{0.025} s_p \sqrt{\frac{1}{n_1} + \frac{1}{n_2} + \frac{1}{n_3}} = 29 \pm 2.052(0.2427)\sqrt{\frac{1}{10} + \frac{1}{8} + \frac{1}{12}} = (28.7235, 29.2765).$$

7.55 From the data, we calculate $\bar{x} = \dfrac{10.7}{3} = 3.5667$ and $s = \left(\dfrac{(65.09 - 10.7)^2}{2}\right) = $

3.6692. A 90% prediction interval for the LC50 of the next test with Diazinon is (degrees of freedom = 2)

$$\bar{x} \pm t_{0.05} s \sqrt{1 + \frac{1}{n}} = 3.5667 \pm 2.920(3.6692)\sqrt{1 + \frac{1}{3}} = (-8.8048, 15.9382).$$

Since LC50 measurements are nonnegative, the interval reduces to $(0, 15.9382)$.

7.57 We calculate $\bar{x} = \dfrac{25.5}{6} = 4.25$ and $s = \left(\dfrac{110.25 - \dfrac{(25.5)^2}{6}}{5}\right)^{1/2} = 0.6124$. A 90%

prediction interval for the compressive strength is (degrees of freedom = 5)

$$\bar{x} \pm t_{0.05} s \sqrt{1 + \frac{1}{n}} = 4.25 \pm 2.015(0.6124)\sqrt{1 + \frac{1}{6}} = (2.9171, 5.5829).$$

7.59 $\bar{x} = 10.1, \quad s = 0.02, \quad n = 100$

a. From table for tolerance limits, K = 1.874.

$$10.1 \pm 1.874(0.02) = (10.0625, 10.1375)$$

b. From table for tolerance limits, K = 1.977.

    $10.1 \pm 1.977(0.02) = (10.0605, 10.1395)$

c. From table for tolerance limits, K = 2.934.

    $10.1 \pm 2.934(0.02) = (10.0413, 10.1587)$

7.61 $\bar{x} = 1.1, s = 0.03, n = 60$. From table for tolerance limits, K = 1.958.

    $1.1 \pm 1.958(0.03) = (1.0413, 1.1587)$

7.63

   a. From table for tolerance limits, K = 1.874.

    $1.1 \pm 1.874(0.03) = (1.0438, 1.1562)$

   b. From table for tolerance limits, K = 1.798.

    $1.1 \pm 1.798(0.03) = (1.0461, 1.1539)$

   Since the K values approach the Z values as n increases, the length of the tolerance interval is not substantially changed for large n values.

7.65

	n	Confidence coefficient
(a)	77	0.903
(b)	18	0.901
(c)	93	0.950
(d)	22	0.952

7.67   A large sample $100(1 - \alpha)\%$ confidence interval for λ is $\hat{\lambda} \pm z_{\alpha/2}$ standard error

$(\hat{\lambda}) = \bar{x} \pm z_{\alpha/2}\sqrt{\dfrac{\lambda}{n}}$. However, since the standard error $(\hat{\lambda})$ depends on λ, which is

unknown, we use the estimate $\hat{\lambda} = \bar{x}$ to get the approximate confidence interval

    $\hat{\lambda} \pm z_{\alpha/2}$ (estimated standard error$(\hat{\lambda})) = \bar{x} \pm z_{\alpha/2}\sqrt{\dfrac{\bar{x}}{n}}.$

b. Assuming that the number of cracks has a Poisson distribution, an approximate 95% confidence interval for the true mean number of cracks per truss is

$$\bar{x} \pm z_{\alpha/2}\sqrt{\frac{\bar{x}}{n}} = 4 \pm 1.96\sqrt{\frac{4}{100}} = (3.608, \quad 4.392)$$

7.69

$$L(\beta) = \prod_{i=1}^{n} f(x_i) = \prod_{i=1}^{n} \frac{x_i^{\alpha-1}e^{-x_i/\beta}}{\beta^{\alpha}\Gamma(\alpha)} = \beta^{-n\alpha}(\Gamma(\alpha))^{-n}e^{-\Sigma(x_i/\beta)\Pi_{i=1}^{n}x_i^{\alpha-1}}$$

To simplify the calculations, we maximize

$$\ln L(\beta) = -n\ln\Gamma(\alpha) + (\alpha-1)\sum_{i=1}^{n}\ln(x_i) - n\alpha\ln\beta - \sum_{i=1}^{n}\frac{x_i}{\beta}.$$

Thus,

$$\frac{\partial \ln L(\beta)}{\partial \beta}\bigg|_{\beta=\hat{\beta}} = -\frac{n\alpha}{\hat{\beta}} + \frac{\sum_{i=1}^{n}x_i}{\hat{\beta}^2} = 0.$$

Solving for $\hat{\beta}$ results in $\hat{\beta} = \dfrac{\sum_{i=1}^{n}x_i}{n\alpha} = \dfrac{\bar{x}}{\alpha}.$

7.71   We need to find the maximum likelihood estimator of p from a geometric (p) distribution.

$$L(p) = \prod_{i=1}^{n} p(1-p)^{x_i-1} = p^{n}(1-p)^{\Sigma_{i=1}^{n}x_i-n}.$$

To simplify the calculations, we equivalently maximize

$$\ln L(p) = n\ln p + \left(\sum_{i=1}^{n}x_i - n\right)\ln(1-p).$$

Then,

$$\frac{\partial \ln L(p)}{\partial p}\bigg|_{p=\hat{p}} = \frac{n}{\hat{p}} - \frac{\left(\sum_{i=1}^{n}x_i - n\right)}{1-\hat{p}} = 0.$$

Solving for $\hat{p}$ gives $\hat{p} = \dfrac{1}{\bar{x}}.$

7.73 First, we compute

$$E(C) = 3E(X) + E(X^2) = 3np + \left(V(X) + (E(X))^2\right)$$

$$= 3np + np(1-p) + n^2p^2 = 4np - np^2 + n^2p^2.$$

For notational convenience, set $E(C) = g(p)$. Thus, the maximum likelihood estimator of $g(p)$ is $g(\hat{p}) = g\left(\dfrac{y}{n}\right)$ and $y =$ number of improperly soldered connections in a microchip with 20 connections. Therefore, for $n = 20$, the maximum likelihood estimator of E(C) is

$$4n\frac{y}{n} - n\left(\frac{y}{n}\right)^2 + n^2\left(\frac{y}{n}\right)^2 = 4y - \frac{y^2}{n} + y^2 = y\left(4 + y - \frac{y}{n}\right) = y\left(4 + y - \frac{y}{20}\right).$$

7.75 Starting with

$$f(y|\lambda) = e^{-\lambda}\frac{\lambda^y}{y!}, \quad y = 0, 1, 2, \ldots$$

Then,

$$f(y|\lambda) = f(y|\lambda)g(\lambda) = e^{-\lambda}\frac{\lambda^y}{y!}e^{-\lambda} = e^{-2\lambda}\frac{\lambda^y}{y!}, \qquad \begin{array}{l} y = 0, 1, 2, \ldots \\ \lambda = 0 \end{array}$$

Now,

$$f(y) = \int_0^\infty f(y, \lambda)d\lambda = \frac{1}{y!}\int_0^\infty e^{-2\lambda}d\lambda = \left(\frac{1}{2}\right)^{y+1}\int_0^\infty \frac{\lambda^{(y+1)-1}e^{-2\lambda}}{\left(\dfrac{1}{2}\right)^{y+1}\Gamma(y+1)}d\lambda$$

$$= \left(\frac{1}{2}\right)^{y+1}, \quad y = 0, 1, 2, \ldots$$

It follows that

$$f(\lambda|y) = \frac{f(y, \lambda)}{f(y)} = \frac{e^{-2\lambda}\lambda^{(y+1)-1}}{y!\left(\dfrac{1}{2}\right)^{y+1}}, \quad \lambda > 0, \quad y = 0, 1, 2, \ldots$$

so that $f(\lambda|y)$ is gamma $\left(\alpha = y + 1, \quad \beta = \dfrac{1}{2}\right)$. Thus, the Bayes estimator is

$$E(\lambda|y) = \alpha\beta = \frac{y+1}{2}.$$

7.77 Assume that the diameter measurements are independent and have a common normal distribution. Then a 90% confidence interval for the mean diameter is (degrees of freedom = 9)

$$\bar{x} + t_{0.05}s/\sqrt{n} = 2.1 \pm \frac{1.833(0.3)}{\sqrt{10}} = (1.9261, \quad 2.2739).$$

7.79 A 95% confidence interval for the mean hardness is (degrees of freedom = 14)

$$\bar{x} \pm \frac{t_{0.025}s}{\sqrt{n}} = 65 \pm \frac{2.145\sqrt{90}}{\sqrt{15}} = (59.7458, 70.2542).$$

7.81

   a. We compute $\bar{x} = \dfrac{23.85}{10} = 2.385$ and the standard deviation

$$s = \left(\frac{58.323 - \dfrac{(23.85)^2}{10}}{9} \right)^{\frac{1}{2}}$$

$= 0.4001$. Assuming a normal population, a 95% confidence interval for the mean pre-etch window width is (degrees of freedom = 9)

$$\bar{x} \pm \frac{t_{0.025}s}{\sqrt{n}} = 2.385 \pm \frac{2.262(0.4001)}{\sqrt{10}} = (2.0988, \quad 2.6712).$$

   b. We compute $\bar{x} = \dfrac{34.26}{10} = 3.425$ and the standard deviation

$$s = \left(\frac{120.5412 - \dfrac{(34.26)^2}{10}}{9} \right)^{\frac{1}{2}}$$

$= 0.5931$. Assuming a normal population, a 95% confidence interval for the mean post-etch window width is

$$\bar{x} \pm \frac{t_{0.025}s}{\sqrt{n}} = \frac{3.426 \pm 2.262(0.5931)}{\sqrt{10}} = (3.0018, 3.8502).$$

7.83

   a. $\left(\dfrac{(n-1)s^2}{\chi^2_{0.025}(n-1)}, \quad \dfrac{(n-1)s^2}{\chi^2_{0.975}(n-1)} \right) = \left(\dfrac{9(0.5931)^2}{19.0228}, \quad \dfrac{9(0.5931)^2}{2.70039} \right) = (0.1664, \quad 1.1724)$

b. $\left(\dfrac{s_2^2}{s_1^2}\dfrac{1}{F_{0.05}(9,9)},\dfrac{s_2^2 1}{s_1^2}F_{0.05}(9,9)\right) = \left(\dfrac{(0.5931)^2}{(0.4001)^2}\dfrac{1}{3.18},\dfrac{(0.5931)^2}{(0.4001)^2}\ 3.18\right)$

$\qquad = (0.6910,\ 6.9879)$

c. Assume both samples are independent and observations come from normal populations.

7.85

a. We calculate $s = \left(\dfrac{2.6779 - \dfrac{(5.17)^2}{10}}{9}\right)^{1/2} = 0.02359$. Then a 95% confidence

interval for the population variance is

$$\left(\dfrac{(n-1)s^2}{\chi^2_{0.025}(n-1)},\dfrac{(n-1)s^2}{\chi^2_{0.975}(n-1)}\right) = \left(\dfrac{9(0.02359)^2}{19.0228},\dfrac{9(0.02359)^2}{2.70039}\right)$$

$\qquad = (0.000263,\ 0.001855).$

b. Assume that weight proportions are normally distributed and the samples are independent. The independence of the samples and the normality of the weight proportions need to be verified.

7.87 Let m be the number of observations allocated to sample 1 and let $n - m$ be the number allocated to sample 2. We need to minimize the length of the confidence interval from $\mu_1 - \mu_2$ given by

$$2z_{\alpha/2}\sqrt{\dfrac{\sigma_1^2}{m} + \dfrac{\sigma_2^2}{n-m}}.$$

Equivalently, we will choose m to minimize $V(\bar{x}_1 - \bar{x}_2) = \dfrac{\sigma_1^2}{m} + \dfrac{\sigma_2^2}{n-m}$. Taking the derivative of V with respect to m, we get

$$\dfrac{dV}{dm} = \dfrac{\sigma_1^2}{m^2} + \dfrac{\sigma_2^2}{m^2} + \dfrac{\sigma_2^2}{(n-m)^2} = 0.$$

Rearranging, we have $\sigma_1^2(n-m)^2 = m^2\sigma_2^2$ so that $m = \dfrac{n\sigma_1}{\sigma_1 + \sigma_2}$ and hence

$$n - m = \dfrac{n\sigma_2}{\sigma_1 + \sigma_2}.$$

7.89 A point estimate of $2\mu_1 + \mu_2$ is $2\bar{y} + \bar{x}$. Then $V(2\bar{y} + \bar{x}) = 4V(\bar{y}) + V(\bar{x})$

$$= 4\frac{\sigma^2}{n} + \frac{3\sigma^2}{m} = \sigma^2\left(\frac{4}{n} + \frac{3}{m}\right).$$ An estimate of the common variance σ^2 is

given by

$$s_p^2 = \frac{(n-1)s_y^2 + \dfrac{(m-1)s_x^2}{3}}{n+m-2}.$$

A 95% confidence interval for $2\mu_1 + \mu_2$ is then

$$2\bar{y} + \bar{x} \pm t_{\alpha/2}s_p\sqrt{\frac{4}{n} + \frac{3}{m}}$$

where s_p is the square root of s_p^2 defined above and degrees of freedom $= n + m - 2$.

7.91 The number of defectives, X, is distributed binomial $(n, \quad p)$. The maximum likelihood estimate of p for fixed n is $\hat{p} = \dfrac{x}{n}$. We seek the maximum likelihood estimate

for $r = \dfrac{\text{number of defectives}}{\text{number of good items}}$. Dividing numerator and denominator of r by the

total number of items gives $r = \dfrac{p}{1-p}$. Hence $r = g(p)$ is a function of p. The

maximum likelihood estimate of r is then given by

$$\hat{r} = g(\hat{p}) = \frac{\dfrac{x}{n}}{1 - \dfrac{x}{n}} = \frac{x}{n-x}.$$

Chapter 8.

8.1 Hypotheses: H_0: $\mu = 130$ $\quad$ H_a: $\mu < 130$

$\quad$ Test Statistics: $\quad z = \dfrac{\bar{x} - \mu_0}{\dfrac{\sigma}{\sqrt{n}}} \approx \dfrac{128.6 - 130}{\dfrac{2.1}{\sqrt{40}}} = -4.22$

$\quad$ Rejection Region: $\quad z < -z_{0.05} = -1.645$

$\quad$ Conclusion: $\quad$ Reject H_0 at $\alpha = 0.05$; i.e., there is sufficient evidence to conclude that the mean output voltage is less than 130 at $\alpha = 0.05$.

8.3 $\quad n \geq \dfrac{(z_{0.05} + z_{0.01})^2 \sigma^2}{(\mu_a - \mu_n)^2} = \dfrac{(1.645 + 2.33)^2 (2.1)^2}{(129 - 130)^2} = 69.68$ or $n = 70$

8.5 $\quad \beta = P(\text{fail to reject } H_0 \text{ given that } H_a \text{ is true})$

$\quad = P\left(\dfrac{\bar{x} - 64}{8/\sqrt{50}} > -2.33 | \mu = 60 \right)$

$\quad = P(\bar{x} > 61.3639 | \mu = 60)$

$\quad = P\left(\dfrac{\bar{x} - 60}{8/\sqrt{50}} > \dfrac{61.3639 - 60}{8/\sqrt{50}} \right)$

$\quad = P(Z > 1.21) = 0.5 - 0.3869 = 0.1131$

8.7 Hypotheses: $\quad H_0 : \mu = 7$ $\qquad H_a : \mu \neq 7$

$\quad$ Test Statistics: $\quad z = \dfrac{\bar{x} - \mu_0}{\sigma/\sqrt{n}} \approx \dfrac{6.8 - 7}{0.9/\sqrt{30}} = -1.22$

$\quad$ Rejection Region: $\quad |z| > z_{0.025} = 1.96$

$\quad$ Conclusion: $\quad$ Fail to reject H_0 at $\alpha = 0.05$; i.e., there is insufficient evidence to conclude that the mean pH is significantly different from 7 at $\alpha = 0.05$.

$\quad$ P-value: $\quad P(|Z| \geq 1.22) = 2(P(Z \geq 1.22)) = 2(0.5 - 0.3888) = 0.2224$

8.9 Hypotheses: $\quad H_0 : p \geq 0.9$ $\qquad H_a : p < 0.9$

$\quad$ Test Statistics: $\quad z = \dfrac{\hat{p} - p_0}{\sqrt{p_0(1 - p_0)/n}} = \dfrac{35/40 - 0.9}{\sqrt{(0.9)(0.1)/40}} = -0.53$

Rejection Region: $z < -z_{0.01} = -2.33$

Conclusion: Fail to reject H_0 at $\alpha = 0.01$; i.e., there is insufficient evidence to conclude that the specification is not being met at $\alpha = 0.01$.

8.11 Summary Statistics: $\bar{x} = 14,510/6 = 2,418.33$

$$s = \left((35,121,500 - (14,510)^2/6)/5\right)^{1/2} = 79.3515$$

Hypotheses: $H_0 : \mu = 2500$ $\qquad$ $H_a : \mu < 2500$

Test Statistics: Assuming a normal population, $t = \dfrac{\bar{x} - \mu_0}{s/\sqrt{n}} = \dfrac{2,418.33 - 2,500}{79.3515/\sqrt{6}}$

$= -2.52$

Rejection Region: $t < -t_{0.01} = -3.365$ (degrees of freedom $= 5$)

Conclusion: Fail to reject H_0 at $\alpha = 0.01$; i.e., there is insufficient evidence to conclude that the mean range of the rockets is less than 2500 after storage.

8.13 Hypotheses: $H_0 : \mu = 30$ $\qquad$ $H_a : \mu \neq 30$

Test Statistics: Assuming that the stress resistance measurements are normally distributed,

$$t = \frac{\bar{x} - \mu_0}{s/\sqrt{n}} = \frac{27.4 - 30}{1.1/\sqrt{10}} = -7.47$$

Rejection Region: $|t| > t_{0.025} = 2.262$ (degrees of freedom $= 9$)

Conclusion: Reject H_0 at $\alpha = 0.05$; i.e., there is sufficient evidence to doubt the specification for stress resistence of the plastic at $\alpha = 0.05$.

8.15 Summary Statistics: $\bar{x} = 34.26/10 = 3.426$

$$s = \left((120.5412 - (34.26)^2/10)/9\right)^{1/2} = 0.5931$$

Hypotheses: $H_0 : \mu = 3.5$ $\qquad$ $H_a : \mu \neq 3.5$

Test Statistics: Assuming the post-etch window widths are normally distributed, $t = \dfrac{\bar{x} - \mu_0}{s/\sqrt{n}} = \dfrac{3.426 - 3.5}{0.5931/\sqrt{10}} = -0.3945$

Rejection Region: $|t| > -t_{0.025} = 2.262$ $\quad$ (degrees of freedom $= 9$)

Conclusion: Fail to reject H_0 at $\alpha = 0.05$; i.e., there is insufficient evidence to conclude that the specifications are being violated at $\alpha = 0.05$.

8.17　Hypotheses:　$H_0 : p \le 0.5$　　$H_a : p > 0.5$

Test Statistics:　$z = \dfrac{\hat{p} - p_0}{\sqrt{p_0(1 - p_0)/n}} = \dfrac{0.53 - 0.5}{\sqrt{(0.5)(0.5)/871}} = 1.77$

P-value:　$P(Z \ge 1.77) = 0.5 - 0.4616 = 0.0384$

Conclusion:　Since the P-value is small (i.e., less than, say, 0.05), we will reject H_0 with a P-value of 0.0384 and conclude that a majority of adults in Florida favor strong support of Israel.

8.19　Hypotheses:　$H_0 : \mu = 300,000$　　$H_a : \mu \neq 300,000$

Test Statistics:　$z = \dfrac{\bar{x} - \mu_0}{\sigma/\sqrt{n}} \approx \dfrac{295,000 - 300,000}{10,000/\sqrt{40}} = -3.16$

Rejection Region:　$|z| > z_{0.05} = 1.645$

Conclusion:　Reject H_0 at $\alpha = 0.1$; i.e., there is sufficient evidence to conclude that the mean tensile strength of the wire fails to meet the specifications.

8.21　Summary Statistics:　$\bar{x} = 57.5/9 = 6.3889$

$$s = \left((368.93 - (57.5)^2/9)/8\right)^{1/2} = 0.4428$$

Hypotheses:　$H_0 : \mu = 6.5$　　$H_a : \mu \neq 6.5$

Test Statistics:　Assuming the pH measurements are normally distributed,

$t = \dfrac{\bar{x} - \mu_0}{s/\sqrt{n}} = \dfrac{6.3889 - 6.5}{0.4428/\sqrt{9}} = -0.753$

Rejection Region:　$|t| > t_{0.025} = 2.306$ (degrees of freedom $= 8$)

Conclusion:　Fail to reject H_0 at $\alpha = 0.05$; i.e., there is insufficient evidence to conclude the mean pH is different from the claimed value of 6.5 at $\alpha = 0.05$.

8.23　Hypotheses:　$H_0 : \sigma^2 \le 100$　　$H_a : \sigma^2 > 100$

Test Statistics:　$\chi^2 = \dfrac{(n-1)s^2}{\sigma_0^2} = \dfrac{14(12)^2}{100} = 20.16$

Rejection Region:　$\chi^2 > \chi^2_{0.05} = 23.6848$ (degrees of freedom $= 14$)

Conclusion:　Fail to reject H_0 at $\alpha = 0.05$; i.e., there is insufficient evidence to conclude that the standard deviation of haul times exceeds the claimed value of 10 minutes at $\alpha = 0.05$.

8.25 Hypotheses: $\quad H_0 : \sigma^2 \leq 400 \qquad H_a : \sigma^2 > 400$

Test Statistics: Assuming a normal population of ranges, $\chi^2 = \dfrac{(n-1)s^2}{\sigma_0^2}$

$$= \frac{5(79.3515)^2}{400} = 78.71.$$

Rejection Region: $\quad \chi^2 > \chi^2_{0.05} = 11.0705$ (degrees of freedom $= 5$)

Conclusion: Reject H_0 at $\alpha = 0.05$; i.e., there is strong evidence that storage significantly increases the variability of the ranges at $\alpha = 0.05$.

8.27 Hypotheses: $\quad H_0 : \mu_1 - \mu_2 = 0 \qquad H_a : \mu_1 - \mu_2 \neq 0$

Test Statistics: $\quad z = \dfrac{\bar{x}_1 - \bar{x}_2}{\sqrt{\sigma_1^2/n_1 + \sigma_2^2/n_2}} = \dfrac{1.65 - 1.43}{\sqrt{(0.26)^2/30 + (0.22)^2/35}} = 3.65$

Rejection Region: $\quad |z| > z_{0.005} = 2.575$

Conclusion: Reject H_0 at $\alpha = 0.01$; i.e., there is sufficient evidence to conclude that the soils significantly differ with respect to the mean shear strength at $\alpha = 0.01$.

8.29 Hypotheses: $\quad H_0 : \mu_1 - \mu_2 = 0 \qquad H_a : \mu_1 - \mu_2 \neq 0$

Test Statistics: Assume independent samples and a common normal distribution.

$$s_p = \left(\frac{6(210)^2 + 9(190)^2}{7 + 10 - 2} \right)^{1/2} = 198.2423$$

$$t = \frac{3520 - 3240}{198.2423\sqrt{1/7 + 1/10}} = 0.10$$

Rejection Region: $\quad |t| > t_{0.025} = 2.131$ (degrees of freedom $= 15$)

Conclusion: Fail to reject H_0 at $\alpha = 0.05$; i.e., there is insufficient evidence to conclude that the methods produce concrete with significantly different mean strengths at $\alpha = 0.05$.

8.31 Summary Statistics:

	1: Natives	2: Nonnatives
n	10	10
$\bar{x}$	87	77.3
s	3.2318	4.0291

Hypotheses: $H_0 : \mu_1 - \mu_2 = 0$ $H_a : \mu_1 - \mu_2 > 0$

Test Statistics: Assume independent samples and a common normal population.

$$s_p = \left(\frac{9(3.2318)^2 + 9(4.0291)^2}{10 + 10 - 2} \right)^{1/2} = 3.6522$$

$$t = \frac{87 - 77.3}{3.6522\sqrt{1/10 + 1/10}} = 5.94$$

Rejection Region: $t < -t_{0.05} = 1.734$ (degrees of freedom = 18)

Conclusion: Reject H_0 at $\alpha = 0.05$; i.e., there is sufficient evidence to conclude that the nonnative English speakers have a significantly smaller mean percentage of correct responses at $\alpha = 0.05$.

8.33   Hypotheses: $H_0 : \sigma_1^2 = \sigma_2^2$ $H_a : \sigma_1^2 \neq \sigma_2^2$

Test Statistics: Assuming normal populations, $F = s_2^2/s_1^2 = (16)^2/(15.3)^2$
$= 1.09$

Rejection Region: $F < 1/F_{0.05}(5,5) = 1/5.05 = 0.1980$ or $F > F_{0.05}(5,5)$
$= 5.05$

Conclusion: Fail to reject H_0 at $\alpha = 0.10$; i.e., there is insufficient evidence to conclude that the variances significantly differ at $\alpha = 0.10$.

8.35   Hypotheses: $H_0 : \mu_1 - \mu_2 = 0$ $H_a : \mu_1 - \mu_2 \neq 0$ where '1' indicates 'original printer' and '2' indicates 'modified printer'

Test Statistics: Assume a common normal population and independent samples.

$$s_p = \left(\frac{9(36) + 7(64)}{10 + 8 - 2} \right)^{1/2} = 6.9462$$

$$t = \frac{98 - 94}{6.9462\sqrt{1/10 + 1/8}} = 1.214$$

The assumption of normality needs to be checked. For example, the exponential distribution may be more appropriate.

Rejection Region: $t > t_{0.01} = 2.583$ (degrees of freedom = $10 + 8 - 2 = 16$)

Conclusion: Fail to reject H_0 at $\alpha = 0.01$; i.e., there is insufficient evidence to conclude that the mean time between failures (MTBF) for the modified printer is less than the MTBF for the original printer at $\alpha = 0.01$.

8.37   Hypotheses: $H_0 : \mu_1 - \mu_2 = 0$ $\qquad$ $H_a : \mu_1 - \mu_2 > 0$

Test Statistics: With $s_p = \left((9(0.03)^2 + 9(0.02)^2)/18\right)^{1/2} = 0.02550$ and assuming a common normal distribution for the resistances,

$$t = \frac{0.19 - 0.11}{0.02550\sqrt{\dfrac{1}{10} + \dfrac{1}{10}}} = 7.015.$$

Rejection Region: $t > t_{0.10} = 1.330$ $\quad$ (d.f. $= 10 + 10 - 2 = 18$)

Conclusion: Reject H_0 at $\alpha = 0.10$; i.e., there is sufficient evidence to conclude that alloying significantly reduces the mean resistance in the wire at $\alpha = 0.10$.

8.39   Summary Statistics: 1: first group of rackets; 2: second group of rackets

$n_1 = 6, \bar{x}_1 = 2,418.33, s_1 = 79.3515; n_2 = 6, \bar{x}_2 = 2,368.33, s_2 = 76.7898;$
$s_p = 78.0812.$

Hypotheses: $H_0 : \mu_1 - \mu_2 = 0$ $\qquad$ $H_a : \mu_1 - \mu_2 \neq 0$

Test Statistics: $t = \dfrac{2418.33 - 2368.33}{78.0812\sqrt{\dfrac{1}{6} + \dfrac{1}{6}}} = 1.109$

Rejection Region: $|t| > t_{0.025} = 2.228$ $\quad$ (degrees of freedom $= 6 + 6 - 2 = 10$)

Conclusion: Fail to reject H_0 at $\alpha = 0.05$; i.e. there is insufficient evidence to conclude that the storage methods produce significantly different mean ranges at $\alpha = 0.05$.

8.41   Since the brands of gasoline are common to both automobiles, we block on the brands and analyze the experiment as a paired sample experiment. We compute the differences for auto A minus auto B: -0.9, -1, 0.9, 0.7, -0.2.

Summary Statistics: $\bar{x}_D = -0.1,$ $\qquad$ $s_D = 0.8803,$ $\qquad$ $n = 5$

Hypotheses: $H_0 : \mu_D = 0$ $\qquad$ $H_a : \mu_D \neq 0$

Test Statistics: Assuming a normal population of differences in gas mileage,

$$t = \frac{-0.1}{0.8803/\sqrt{5}} = -0.254.$$

Rejection Region : $|t| > t_{0.025} = 2.776$ (degrees of freedom $= 4$)

Conclusion: Fail to reject H_0 at $\alpha = 0.05$; i.e., there is insufficient evidence to conclude a significant difference between mean mileage figures for the two automobiles at $\alpha = 0.05$.

8.43 Since the type of powder is common to both procedures, we block on powder type and analyze the experiment as a paired sample experiment. The differences for procedure I minus procedure II are: $-2, 1, -3, -2, 1, 3$.

Summary Statistics: $\bar{x}_D = -1/3,$ $s_D = 2.3381,$ $n = 6$

Hypotheses: $H_0 : \mu_D = 0$ $H_a : \mu_D \neq 0$

Test Statistics: Assuming a normal population for the porosity differences,

$$t = \frac{-0.3333}{2.3381/\sqrt{6}} = -0.349.$$

Rejection Region: $|t| > t_{0.025} = 2.571$ (degrees of freedom $= 5$)

Conclusion: Fail to reject H_0 at $\alpha = 0.05$; i.e., there is insufficient evidence to conclude that the mean porosities of the two procedures significantly differ at $\alpha = 0.05$.

8.45 Hypotheses: $H_0 : \sigma_1^2 = \sigma_2^2$ $H_a : \sigma_1^2 \neq \sigma_2^2$

Test Statistics: Assuming normal populations,

$$F = s_1^2/s_2^2 = (79.3515)^2/(76.7898)^2 = 1.07.$$

Rejection Region: $F > F_{0.05}(5,5) = 5.05$ or $F < 1/F_{0.05}(5,5) = 1/5.05$
$= 0.198$

Conclusion: Fail to reject H_0 at $\alpha = 0.10$; i.e., there is insufficient evidence to conclude that the variance among range measurements significantly differs for the two storage methods at $\alpha = 0.10$.

8.47 Hypotheses: $H_0: p_1 = \dfrac{5}{10}, \quad p_2 = \dfrac{2}{10}, \quad p_3 = \dfrac{2}{10}, \quad p_4 = \dfrac{1}{10}$

H_a: At least one equality fails to hold

Test Statistics: Observed and expected cell counts are listed below.

					Totals
n_i	48	18	21	13	100
$E(n_i)$	50	20	20	10	100

$$\chi^2 = \sum_{i=1}^{4} \frac{(n_i - E(n_i))^2}{E(n_i)} = \frac{(48-50)^2}{50} + \cdots + \frac{(13-10)^2}{10} = 1.23$$

Rejection Region: $\chi^2 > \chi^2_{0.05} = 7.81$ (degrees of freedom = 3)

Conclusion: Fail to reject H_0 at $\alpha = 0.05$; i.e., there is insufficient evidence to conclude that the proportions differ from the null hypothesis at $\alpha = 0.05$.

8.49 Hypotheses: $H_0: p_1 = p_2 = p_3 \quad H_a$: At least one inequality holds

Test Statistics: Observed and expected (values in parentheses) cell counts are listed in the table below.

	I	II	III	Totals
Defectives	6(6.67)	5(6.67)	9(6.67)	20
Nondefectives	24(23.3)	25(23.3)	21(23.3)	70
Totals	30	30	30	90

$$\chi^2 = \sum_{i=1}^{3} \frac{(Y_i - E(Y_i))^2}{E(Y_i)} + \frac{((n_i - Y_i) - E(n_i - Y_i))^2}{E(n_i - Y_i)}$$

$$= \frac{(6-6.67)^2}{6.67} + \cdots + \frac{(21-23.3)^2}{23.3} = 1.67$$

Rejection Region: $\chi^2 > \chi^2_{0.05} = 5.99147$

(degrees of freedom = $(2-1)(3-1) = 2$)

Conclusion: Fail to reject H_0 at $\alpha = 0.05$; i.e., there is insufficient evidence to conclude that the lines significantly differ with respect to the number of defectives produced at $\alpha = 0.05$.

8.51 Hypotheses: $H_0 : p_1 = p_2$ $H_a : p_1 \neq p_2$

Test Statistics: Observed and expected (values in parentheses) cell frequencies are listed in the following table.

	A	B	Totals
Infected	20(18)	16(18)	36
Noninfected	80(82)	84(82)	164
Total	100	100	200

$$\chi^2 = \frac{(20-18)^2}{18} + \cdots + \frac{(84-82)^2}{82} = 0.54$$

Rejection Region: $\chi^2 > \chi^2_{0.01} = 6.63490$

(degrees of freedom $= (2-1)(2-1) = 1$)

Conclusion: Fail to reject H_0 at $\alpha = 0.01$, i.e. the chemicals do not significantly differ in their ability to protect trees at $\alpha = 0.01$.

8.53

a. Hypotheses: H_0: Athletic involvement and GPA are independent

H_a: Athletic involvement and GPA are not independent

Test Statistics: Observed and expected (values in parentheses) cell frequencies are given in the following table.

Athletic Involvement (semesters)

GPA	0	1-3	4 or more	Totals
Below mean	290(264)	94(109.5)	42(52.5)	426
Above mean	238(264)	125(109.5)	63(52.5)	426
Totals	528	219	105	852

$$\chi^2 = \frac{(290-264)^2}{264} + \cdots + \frac{(63-52.5)^2}{52.5} = 13.71$$

Rejection Region: $\chi^2 > \chi^2_{0.05} = 5.99147$

(degrees of freedom $= (2-1)(3-1) = 2$)

Conclusion: Reject H_0 at $\alpha = 0.05$; i.e., there is sufficient evidence to conclude that GPA and athletic involvement are not independent at $\alpha = 0.05$.

b. Let $p =$ proportion with grades above the mean and $1 - p =$ proportion with grades below the mean.

Hypotheses: $H_0 : p = 0.5$ $\qquad$ $H_a : p \neq 0.5$

Test Statistics: $z = \dfrac{\hat{p} - p_0}{\sqrt{p_0(1 - p_0)/n}} = \dfrac{63/105 - 0.5}{\sqrt{(0.5)(0.5)/105}} = 2.05$

Rejection Region: $|z| > z_{0.025} = 1.96$

Conclusion: Reject H_0 at $\alpha = 0.05$; i.e., there is sufficient evidence to conclude that the proportion with grades above the mean is significantly different from the proportion below the mean at $\alpha = 0.05$.

8.55

a. Hypotheses: $H_0 : p_1 = p_2 = p_3 = p_4 = 0.25$

H_a: At least one equality does not hold

Test Statistic: Observed and expected frequencies are presented in the following table.

	1	2	3	4	Totals
n_i	42	36	31	41	150
$E(n_i)$	37.5	37.5	37.5	37.5	150

$$\chi^2 = \frac{(42 - 37.5)^2}{37.5} + \cdots + \frac{(41 - 37.5)^2}{37.5} = 2.05$$

Rejection Region: $\chi^2 > \chi^2_{0.05} = 7.81473$ (degrees of freedom $= 3$)

Conclusion: Fail to reject H_0 at $\alpha = 0.05$; i.e., there is insufficient evidence to conclude that the proportions differ at $\alpha = 0.05$.

b. Let $p = $ P(subway entrances 1 and 2 are used) $= 1 - $ P(ground-level entrances 3 and 4 are used).

Hypotheses: $H_0 : p = 0.5$ $H_a : p \neq 0.5$

Test Statistics: $z = \dfrac{\hat{p} - p_0}{\sqrt{p_0(1 - p_0)/n}} = \dfrac{(42 + 36)/150 - 0.5}{\sqrt{(0.5)(0.5)/150}} = 0.49$

Rejection Region: $|z| > z_{0.025} = 1.96$

Conclusion: Fail to reject H_0 at $\alpha = 0.05$; i.e., there is insufficient evidence to conclude that the proportions using subway and ground-level entrances differ at $\alpha = 0.05$.

8.57   Hypotheses: $H_0 : p_1 = p_2$

$H_a : p_1 \neq p_2$ where '1' indicates '1983' and '2' indicates '1984'

Test Statistics: Observed and expected (values in parentheses) cell frequencies are presented in the following table.

	1983	1984	Totals
Fear effects on ozone	105(96)	87(96)	192
Do not fear effects on ozone	195(204)	213(204)	408
Totals	300	300	600

$$\chi^2 = \frac{(105 - 96)^2}{96} + \cdots + \frac{(213 - 204)^2}{204} = 2.48$$

Rejection Region: $\chi^2 > \chi^2_{0.05} = 6.63490$

(degrees of freedom $= (2 - 1)(2 - 1) = 1$)

Conclusion: Fail to reject H_0 at $\alpha = 0.01$; i.e., there is insufficient evidence to conclude that the proportions fearing the effects of ozone in 1983 and 1984 differ at $\alpha = 0.01$.

8.59   Hypotheses: $H_0 : p_1 = \cdots = p_7 = 0.012$ H_a: At least one inequality holds

Test Statistics: Observed and expected (values in parentheses) cell frequencies are presented in the following table.

	Failures	Successes
1:0-6	1(0.34)	27(27.66)
2:6-12	3(1.10)	89(90.90)
3:12-18	1(3.24)	269(266.76)
4:18-24	9(8.42)	693(693.58)
5:24-30	10(7.79)	639(641.21)
6:30-36	0(1.61)	134(132.39)
7:>36	0(1.12)	93(91.88)

$$\chi^2 = \frac{(1 - 0.34)^2}{0.34} + \cdots + \frac{(93 - 91.88)^2}{91.88} = 9.62$$

Rejection Region: $\chi^2 > \chi^2_{0.05} = 12.5916$

(degrees of freedom $= (2 - 1)(7 - 1) = 6$)

Conclusion: Fail to reject H_0 at $\alpha = 0.05$; i.e., there is insufficient evidence that the proportions significantly differ at $\alpha = 0.05$.

8.61   Hypotheses: $H_0 : p_1 = p_2$

$H_a : p_1 \neq p_2$ where '1' denotes 'earlier poll' and '2' denotes 'later poll'

Test Statistics: Observed and expected (values in parentheses) cell frequencies are given in the following table.

About Nuclear War	1	2	Totals
Worry a lot	500(550)	600(550)	1100
Don't worry a lot	500(450)	400(450)	900
Total	1000	1000	2000

$$\chi^2 = \frac{(500 - 550)^2}{550} + \cdots + \frac{(400 - 450)^2}{450} = 20.20$$

Rejection Region: $\chi^2 > \chi^2_{0.01} = 6.63490$

(degrees of freedom $= (2 - 1)(2 - 1) = 1$)

Conclusion: Reject H_0 at $\alpha = 0.01$; i.e., there is sufficient evidence to conclude that the proportions 'worrying' about nuclear war significantly differ in the two polls at $\alpha = 0.01$.

8.63 $\hat{\lambda} = \bar{y} = \dfrac{\sum_i y_i F_i}{n} = \dfrac{199}{50} = 3.98$

y_i	F_i	$y_i F_i$	$\hat{p}_i = e^{-\hat{\lambda}} \hat{\lambda}^{y_i}/y_i!$	$\hat{E}(F_i) = n\hat{p}_i$
0	0 ⎤	0	0.01869	0.934 ⎤
	⎬ 3			⎬ 4.652
1	3 ⎦	3	0.07437	3.718 ⎦
2	5	10	0.14799	7.400
3	10	30	0.19634	9.817
4	14	56	0.19536	9.768
5	8	40	0.15550	7.775
6 or more	10	60	0.21175	10.588
	n=50	199		

Hypotheses: H_0: Population has a Poisson distribution $p(y) = e^{-\lambda}\lambda^2/y!$, $y = 0, 1, 2, \ldots$

H_a: Population does not have a Poisson distribution

Test Statistics: $\chi^2 = \dfrac{(3 - 4.652)^2}{4.652} + \dfrac{(5 - 7.400)}{7.400} + \cdots + \dfrac{(10 - 10.588)^2}{10.588}$

$= 3.24$

Rejection Region: $k =$ number of cells $= 6$

$m =$ number of parameters estimated $= 1$

$\chi^2 > \chi^2_{0.05} = 9.48773$ (degrees of freedom $= (k - 1) - m = (6 - 1) - 1 = 4$)

Conclusion: Fail to reject H_0 at $\alpha = 0.05$; i.e., there is insufficient evidence to conclude that the Poisson model is inadequate at $\alpha = 0.05$.

8.65 $\quad \hat{\lambda} = \bar{y} = \dfrac{\sum_i y_i F_i}{n} = \dfrac{200}{414} = 0.4831$

y_i	F_i	$y_i F_i$	$\hat{p}_i = e^{-\hat{\lambda}} \hat{\lambda}^{y_i}/y_i!$	$\hat{E}(F_i) = n\hat{p}_i$
0	296	0	0.61687	255.384
1	74	74	0.29801	123.376
2	26	52	0.07198	29.800
3	8 ⌐	24	0.01159	4.798 ⌐
4	4	16	0.00140	0.580
5	4 ⊢18	20	0.00014	0.058 ⊢5.440
6	1	6	0.00001	0.004
7	0	0	0.00000	0.000
8	1 ⌐	8	0.00000	0.000 ⌐

$\qquad \qquad n = 414 \qquad \qquad 200$

Hypotheses: $\quad H_0$: Population has a Poisson distribution $p(y) = e^{-\lambda}\lambda^y/y!$, $y = 0, 1, 2, \ldots$

H_a: Population does not have Poisson distribution

Test Statistics: $\quad \chi^2 = \dfrac{(296 - 255.384)^2}{255.384} + \cdots + \dfrac{(18 - 5.440)^2}{5.440} = 55.70$

Rejection Region: $\quad k = $ number of cells $= 4$

$m = $ number of parameters estimated $= 1$

$\chi^2 > \chi^2_{0.05} = 5.99147$ (degrees of freedom $= (k-1) - m = (4-1) - 1 = 2$)

Conclusion: $\quad$ Reject H_0 at $\alpha = 0.05$; i.e., there is sufficient evidence to conclude that the Poisson model is inadequate at $\alpha = 0.05$.

8.67 $\quad$ We use the Kolmogorov-Smirnov test. The maximum-likelihood estimates for μ and σ^2 are obtained for $y_i = \ln x_i$ as

$$\hat{\mu} = \bar{y} = \frac{\sum_i y_i}{n} = \frac{13.6732}{15} = -0.9115$$

$$\hat{\sigma}^2 = \frac{n-1}{n} s^2 = \frac{(34.8214 - (-13.6732)^2/15)}{15} = 1.4905.$$

118

x_i	$z_i = \dfrac{y_i - \hat{\mu}}{\hat{\sigma}}$	i/n	$\dfrac{i}{n} - F(z_i)$	$F(z_i) - \dfrac{i-1}{n}$
0.075	-1.3750	0.0667	-0.0171	0.0838
0.100	-1.1394	0.1333	0.0062	0.0604
0.100	-1.1394	0.2000	0.0729	-0.0062
0.150	-0.8073	0.2667	0.0577	0.0090
0.230	-0.4572	0.3333	0.0105	0.0561
0.290	-0.2673	0.4000	0.0064	0.0603
0.310	-0.2127	0.4667	0.0499	0.0168
0.320	-0.1867	0.5333	-0.1086	-0.0420
0.330	-0.1615	0.6000	0.1636	-0.0969
0.460	0.1106	0.6667	0.1229	-0.0562
0.540	0.2419	0.7333	0.1385	-0.0719
0.850	0.6135	0.8000	0.0709	-0.0042
1.300	0.9615	0.8667	0.0352	0.0315
1.900	1.2723	0.9333	0.0353	0.0313
9.000	2.5464	1.0000	0.0055	0.0613

Hypotheses: H_0: Population has a lognormal distribution

H_a: Population does not have a lognormal distribution

Test Statistics: $D = \max(D^+, D^-) = \max(0.1636, 0.0838) = 0.1636$

$$\text{modified } D = 0.1636 \left(\sqrt{15} - 0.01 + \frac{0.85}{\sqrt{15}} \right) = 0.6679$$

Rejection Region: modified $D > 0.895$ (from Table 8.5)

Conclusion: Fail to reject H_0 at $\alpha = 0.05$; i.e., there is insufficient evidence to conclude that the lognormal distribution is inadequate at $\alpha = 0.05$.

8.69   We use the Kolmogorov-Smirnov test. From the logarithms of the data we compute the maximum-likelihood estimates $\hat{\mu} = \bar{x} = 1.8499$ and $\hat{\sigma}^2 = \dfrac{n-1}{n} s^2 = 0.3203$.

Then for $z_{(i)} = \dfrac{\ln(x_i) - \hat{\mu}}{\hat{\sigma}}$ we have

$x_{(i)}$	$z_{(i)}$	i/n	$\dfrac{i}{n} - F(z_i)$	$F(z_i) - \dfrac{i-1}{n}$
2.0	-2.04	0.05	0.0293	0.0207
3.0	-1.33	0.10	0.0082	0.0418
3.1	-1.27	0.15	0.0480	0.0020
4.3	-0.69	0.20	-0.0451	0.0951
4.4	-0.65	0.25	-0.0078	0.0578
4.8	-0.50	0.30	-0.0085	0.0585
4.9	-0.46	0.35	0.0272	0.0228
5.1	-0.39	0.40	0.0517	-0.0017
5.4	-0.29	0.45	0.0641	-0.0141
5.7	-0.19	0.50	0.0753	-0.0253
6.1	-0.07	0.55	0.0779	-0.0279
6.6	0.07	0.60	0.0721	-0.0221
7.3	0.24	0.65	0.0552	-0.0052
7.6	0.32	0.70	0.0745	-0.0245
8.3	0.47	0.75	0.0692	-0.0192
9.1	0.63	0.80	0.0643	-0.0143
11.2	1.00	0.85	0.0087	0.0413
14.4	1.44	0.90	-0.0251	0.0751
16.7	1.71	0.95	-0.0064	0.0564
19.8	2.01	1.00	0.0222	0.0278

8.71   We wish to test the fit of a Weibull ($\gamma = 3$) distribution. Equivalently, we will work with $y_i = x_i^3$ so that y will have an exponential (θ) distribution. The maximum-likelihood estimate of θ is given by $\hat{\theta} = \bar{y} = 46.1698$ and $F(y) = 1 - e^{-y/46.1698}$.

y_i	$F(y_i)$	i/n	$\dfrac{i}{n} - F(y_i)$	$F(y_i) - \dfrac{i-1}{n}$
5.832	0.1187	0.125	0.0063	0.1187
13.824	0.2587	0.250	-0.0087	0.1337
21.952	0.3784	0.375	-0.0034	0.1284
32.768	0.5082	0.500	-0.0082	0.1332
46.656	0.6360	0.625	-0.0110	0.1360
59.319	0.7233	0.750	0.0267	0.0983
85.184	0.8420	0.875	0.0330	0.0920
103.823	0.8945	1.000	0.1055	0.0195

Hypotheses:　H_0: Population has a Weibull $(\gamma = 3)$ distribution

H_a: Population does not have a Weibull $(\gamma = 3)$ distribution

Test Statistics:　$D = \max\,(D^+, D^-) = \max\,(0.1055, 0.1360) = 0.1360$

$$\text{modified } D = \left(0.136 - \frac{0.2}{8}\right)\left(\sqrt{8} + 0.26 + \frac{0.5}{\sqrt{8}}\right) = 0.3624$$

Rejection Region:　modified $D > 1.190$ (from Table 8.5)

Conclusion:　Fail to reject H_0 at $\alpha = 0.025$; i.e., there is insufficient evidence to conclude that the Weibull $(\gamma = 3)$ model is inadequate at $\alpha = 0.025$.

8.73

a.

p	P(A)
0.1	0.930
0.2	0.678
0.3	0.383
0.4	0.167
0.6	0.012

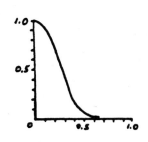

b.

p	P(A)
0.1	0.998
0.3	0.850
0.4	0.633
0.5	0.377
0.6	0.166
0.7	0.047

c.

p	P(A)
0.05	0.925
0.1	0.677
0.2	0.206
0.3	0.035

d.

p	P(A)
0.1	0.957
0.2	0.630
0.3	0.238
0.4	0.051

8.75 From Table 17, a lot size of 1,200 and a level IV plan give a code letter K. The AQL conversion table indicates that we use an AQL value of 2.5. From Table 18, we find

	Sample size	k
(a)	35	1.57
(b)	35	1.76

where we accept the lot if $\dfrac{20 - \bar{x}}{s} \geq k$.

8.77 From Table 18, a lot size of 250 and a level IV plan give a code letter H. From Table 18, we find

	Sample size	k
(a)	20	0.917
(b)	20	1.12

where we accept the lot if $\dfrac{\bar{x} - L}{s} \geq k$.

8.79

    a. Test for positive plates.

        Hypotheses: $H_0 : \mu = 120$ $H_a : \mu \neq 120$

        Test Statistics: Assuming a normal population, $t = \dfrac{111.6 - 120}{2.5/\sqrt{16}} = -13.44$

        P-value: $P(|t| > 13.44) < 0.01$ with degrees of freedom $= 15$

        Conclusion: Since the P-value is small, we have sufficient evidence to reject H_0 and conclude that the positive plates do not meet the thickness specification.

    b. Test for negative plates.

        Hypotheses: $H_0 : \mu = 100$ $H_a : \mu \neq 100$

        Test Statistics: Assuming a normal population, $t = \dfrac{99.44 - 100}{3.7/\sqrt{9}} = -0.454$

        P-value: $P(|t| > 0.454) < 0.2$

        Conclusion: Since the P-value is large, we fail to reject H_0 and conclude that we do not have sufficient evidence to doubt that the negative plates meet the thickness specification.

8.81 Hypotheses: $H_0 : \mu \geq 45$ $H_a : \mu < 45$

        Test Statistics: $z = \dfrac{44 - 45}{3/\sqrt{40}} = -2.11$

        P-value: $P(Z < -2.11) = 0.05 - 0.4826 = 0.0174$

Conclusion: Since the P-value is small (say, less than 0.05), we reject H_0 and conclude that the mean hours worked per week are significantly less than 45. Thus, prospective employees are probably not being told the true mean weekly hours of work.

8.83   Hypotheses: $H_0 : \mu = 10$ $H_a : \mu \neq 10$

Test Statistics: Assuming a normal population, $\bar{x} = 9$, $s = 6.4244$,

$$t = \frac{9 - 10}{6.4244\sqrt{12}} = -0.54.$$

Rejection Region: $|t| > t_{0.025} = 2.201$ (degrees of freedom $= 11$)

Conclusion: Fail to reject H_0 at $\alpha = 0.05$; i.e., there is insufficient evidence that the mean LC50 for DOT is different from 10 parts per million at $\alpha = 0.05$. The data tend to support the standard.

8.85   Hypotheses: $H_0 : \sigma^2 \leq 0.16$ $H_a : \sigma^2 > 0.16$

Test Statistics: Assuming a normal population, $\chi^2 = \dfrac{(15 - 1)(0.5)^2}{0.16} = 21.88.$

Rejection Region: $\chi^2 > \chi^2_{0.01} = 21.0642$ (degrees of freedom $= 14$)

Conclusion: Reject H_0 at $\alpha = 0.10$; i.e., there is sufficient evidence to refute the claim that the standard deviation does not exceed 0.4 ohm at a level of significance $\alpha = 0.10$.

8.87   Hypotheses: $H_0 : \mu_A - \mu_B = 0$ $H_a : \mu_A - \mu_B \neq 0$

Test Statistics: $z = \dfrac{0.18 - 0.21}{\sqrt{(0.02)^2/35 + (0.03)^2/30}} = -4.66$

Rejection Region: $|z| > z_{0.025} = 1.96$

Conclusion: Reject H_0 at $\alpha = 0.05$; i.e., there is sufficient evidence to conclude that the mean pit depths differ for the two types of coating at $\alpha = 0.05$.

8.89   Hypotheses: $H_0 : p_A - p_B = 0$ $H_a : p_A - p_B \neq 0$

Test Statistics: $\hat{p}_A = \dfrac{25}{30} = 0.83$ $\hat{p}_B = \dfrac{25}{35} = 0.71$

$$z = \frac{0.83 - 0.71}{\sqrt{\dfrac{(0.83)(0.17)}{30} + \dfrac{(0.71)(0.29)}{35}}} = 1.17$$

Rejection Region: $|z| > z_{0.025} = 1.96$

Conclusion: Fail to reject H_0 at $\alpha = 0.05$; i.e., there is insufficient evidence to conclude that the proportions voicing no objections differ between the two companies at $\alpha = 0.05$.

8.91   Hypotheses: $H_0 : \sigma_1^2 = \sigma_2^2$ $H_a : \sigma_2^2 > \sigma_1^2$ where '1' denotes 'wood' and '2' denotes 'graphite'

Test Statistics: $F = s_2^2 / s_1^2 = \dfrac{(0.07)^2}{(0.02)^2} = 12.25$

Rejection Region: $F > F_{0.05}(2, 2) = 19.00$

Conclusion: Fail to reject H_0 at $\alpha = 0.05$; i.e., there is insufficient evidence to conclude that the graphite rackets give more variable results than the wood rackets at $\alpha = 0.05$.

8.93   We would expect the subjects to pick their partners correctly 1/3 of the time. Hence

$$H_0 : p = 1/3 \qquad H_a : p > 1/3$$

are appropriate hypotheses.

Chapter 9.

9.1

a. $\sum x_i = 21$ $\sum x_i^2 = 91$ $\sum x_i y_i = 78$

$\sum y_i = 18$ $\sum y_i^2 = 68$ $n = 6$

$SS_{xx} = 91 - \dfrac{(21)^2}{6} = 17.5$ $SS_{yy} = 68 - \dfrac{(18)^2}{6} = 14$

$SS_{xy} = 78 - \dfrac{(21)(18)}{6} = 15$ $\hat{\beta}_1 = \dfrac{SS_{xy}}{SS_{xx}} = \dfrac{15}{17.5} = \dfrac{6}{7} = 0.8571$

$\hat{\beta}_0 = \bar{y} - \hat{\beta}_1 \bar{x} = \dfrac{18}{6} - \left(\dfrac{6}{7}\right)\left(\dfrac{21}{6}\right) = 0$

b.

The least-squares line does not pass through the data points.

9.3 $\sum x_i = 10$ $\sum x_i^2 = 22.5$ $\sum x_i y_i = 61.95$

$\sum y_i = 27.5$ $\sum y_i^2 = 170.77$ $n = 5$

$SS_{xx} = 22.5 - \dfrac{(10)^2}{5} = 2.5$ $SS_{yy} = 170 - \dfrac{(27.5)^2}{5} = 19.52$

$SS_{xy} = 61.95 - \dfrac{(10)(27.5)}{5} = 6.95$ $\hat{\beta}_1 = \dfrac{SS_{xy}}{SS_{xx}} = \dfrac{6.95}{2.5} = 2.78.$

$\hat{\beta}_0 = \bar{y} - \hat{\beta}_1 \bar{x} = \dfrac{27.5}{5} - (2.78)\left(\dfrac{10}{5}\right) = -0.06$

Least-squares line: $\hat{y} = -0.06 + (2.78)x$

9.5 $\sum x_i = 556$ $\sum x_i^2 = 39,080$ $\sum x_i y_i = 120,399$

$\sum y_i = 1,756$ $\sum y_i^2 = 391,720$ $n = 8$

$$SS_{xx} = 39,080 - \frac{(556)^2}{8} = 438 \qquad SS_{yy} = 391,720 - \frac{(1,756)^2}{8} = 6,278$$

$$SS_{xy} = 120,399 - \frac{(556)(1,756)}{8} = -1,643 \qquad \hat{\beta}_1 = \frac{SS_{xy}}{SS_{xx}} = \frac{-1,643}{438} =$$

$$-3.7511$$

$$\hat{\beta}_0 = \bar{y} - \hat{\beta}_1 \bar{x} = \frac{(1,756)}{8} - (-3.7511)\left(\frac{556}{8}\right) = 480.2043$$

Least-squares line: $\hat{y} = 480.2043 - (3.7511)x$

9.7 $\quad \sum x_i = 180 \qquad \sum x_i^2 = 2,900 \qquad \sum x_i y_i = 7,420$

$\qquad \sum y_i = 469 \quad \sum y_i^2 = 19,313 \qquad n = 12$

$$SS_{xx} = 2,900 - \frac{(180)^2}{12} = 200 \qquad SS_{yy} = 19,313 - \frac{(469)^2}{12} = 982.9167$$

$$SS_{xy} = 7,420 - \frac{(180)(469)}{12} = 385 \qquad \hat{\beta}_1 = \frac{SS_{xy}}{SS_{xx}} = \frac{385}{200} = 1.925$$

$$\hat{\beta}_0 = \bar{y} - \hat{\beta}_1 \bar{x} = \frac{(469)}{12} - (1.925)\left(\frac{180}{12}\right) = 10.2083$$

Least-squares line: $\hat{y} = 10.2083 + 1.925x$

9.9 Using the data from 9.1−9.8, we calculate

$$SSE = SS_{yy} - \hat{\beta}_1 SS_{xy} = SS_{yy} - \left(\frac{SS_{xy}}{SS_{xx}}\right) SS_{xy} = SS_{yy} - \frac{(SS_{xy})^2}{SS_{xx}}$$

and $s^2 = \dfrac{SSE}{(n-2)}$.

9.1 $\quad SSE = 14 - \dfrac{(15)^2}{17.5} = 1.1429 \qquad\qquad s^2 = \dfrac{1.1429}{4} = 0.2857$

9.2 $\quad SSE = 16 - \dfrac{(-12)^2}{10} = 1.6 \qquad\qquad s^2 = \dfrac{1.6}{3} = 0.5333$

9.3 $\quad SSE = 19.52 - \dfrac{(6.95)^2}{2.5} = 0.199 \qquad\qquad s^2 = \dfrac{0.199}{3} = 0.0663$

9.4 $\quad SSE = 0.7533 - \dfrac{(3.6667)^2}{18.8333} = 0.03947 \qquad\qquad s^2 = \dfrac{0.03947}{4} = 0.009867$

9.5 $\quad SSE = 6,278 - \dfrac{(-1,643)^2}{438} = 114.8744$ $\qquad s^2 = \dfrac{114.8744}{6} = 19.1457$

9.6 $\quad SSE = 7,639.275 - \dfrac{(2,424.1535)^2}{778.2792} = 88.6164$ $\qquad s^2 = \dfrac{88.6164}{6} = 14.7694$

9.7 $\quad SSE = 982.9167 - \dfrac{(385)^2}{200} = 241.7917$ $\qquad s^2 = \dfrac{241.7917}{10} = 24.1792$

9.8 $\quad SSE = 1,101.1686 - \dfrac{(1,546.553)^2}{2,360.2388} = 87.7855$ $\qquad s^2 = \dfrac{87.7855}{8} = 10.9732$

9.11 Hypotheses: $H_0 : \beta_1 = 0$ $\qquad H_a : \beta_1 \neq 0$

Test Statistics: Assume that errors are independent, normal $(0, \quad \sigma^2)$.

$$t = \frac{1}{s/\sqrt{SS_{xx}}} = \frac{6/7}{\sqrt{0.2857}/\sqrt{17.5}} = 6.71$$

Rejection Region: $|t| > t_{0.025} = 2.776$ $\qquad$ (degrees of freedom $= 6 - 2 = 4$)

Conclusion: Reject H_0 at $\alpha = 0.05$; i.e., there is sufficient evidence to conclude that the slope is significantly different from zero at $\alpha = 0.05$.

9.13 $\quad \hat{\beta}_1 \pm t_{0.025} \dfrac{s}{\sqrt{SS_{xx}}} = 2.78 \pm 3.182 \dfrac{\sqrt{0.0663}}{\sqrt{2.5}} = (2.2618, \quad 3.2982)$

(degrees of freedom $= 5 - 2 = 3$)

9.15 Hypotheses: $H_0 : \beta_1 = 0$ $\qquad H_a : \beta_1 \neq 0$

Test Statistic: Assume that errors are independent, normal $(0, \quad \sigma^2)$.

$$t = \frac{1}{s/\sqrt{SS_{xx}}} = \frac{-3.7511}{\sqrt{19.1457}/\sqrt{438}} = -17.94$$

Rejection Region: $|t| > t_{0.025} = 2.447$ $\qquad$ (degrees of freedom $= 8 - 2 = 6$)

Conclusion: Reject H_0 at $\alpha = 0.05$; i.e., there is a significant linear relationship between Rockwell hardness and abrasion loss at $\alpha = 0.05$.

9.17 We seek a confidence interval for $5\beta_1$. If we assume that the errors are independent,

normal $(0, \quad \sigma^2)$, then $\hat{\beta}_1$ is distributed normal $\left(\beta_1, \quad \dfrac{\sigma^2}{SS_{xx}}\right)$. Hence $5\hat{\beta}_1$ has

mean $E(5\hat{\beta}_1) = 5E(\hat{\beta}_1) = 5\beta_1$ and $V(5\hat{\beta}_1) = 25V(\hat{\beta}_1) = \dfrac{25\sigma^2}{SS_{xx}}$. Then $5\hat{\beta}_1$ is

distributed normal $\left(5\beta_1, \quad \dfrac{25\sigma^2}{SS_{xx}}\right)$. Thus a 95% confidence interval for $5\beta_1$ is

$$5\hat{\beta}_1 \pm t_{0.025}\frac{5s}{\sqrt{SS_{xx}}} = 5(1.925) \pm 2.228(5)\frac{\sqrt{24.1792}}{\sqrt{200}} = (5.752, \quad 13.498)$$

(degrees of freedom $= 12 - 2 = 10$)

9.19 $\sum x_i = 55 \qquad \sum x_i^2 = 385 \qquad \sum x_i y_i = 1,889.09$

$\quad\quad \sum y_i = 348.47 \qquad \sum y_i^2 = 12,153.6443 \qquad n = 10$

$\quad\quad SS_{xx} = 385 - \dfrac{(55)^2}{10} = 82.5 \qquad SS_{yy} = 12,153.6443 - \dfrac{(348.47)^2}{10} = 10.5102$

$\quad\quad SS_{xy} = 1,889.09 - \dfrac{(55)(348.47)}{10} = -27.495$

a. $\hat{\beta}_1 = \dfrac{SS_{xy}}{SS_{xx}} = \dfrac{-27.495}{82.5} = -0.3333$

$\quad\quad \hat{\beta}_0 = \bar{y} - \hat{\beta}_1\bar{x} = \dfrac{348.47}{10} - (-0.3333)\left(\dfrac{55}{10}\right) = 36.68$

Least-squares line: $\bar{y} = 36.68 - 0.3333x$

b. $SSE = \dfrac{SS_{yy} - (SS_{xy})^2}{SS_{xx}} = \dfrac{10.5102 - (-27.495)^2}{82.5} = 1.3469$

$\quad\quad s^2 = \dfrac{SSE}{n-2} = \dfrac{1.3469}{8} = 0.1684$

c. Hypotheses: $H_0 : \beta_1 = 0 \qquad H_a : \beta_1 < 0$

Test Statistics: Assume that errors are independent, normal $(0, \quad \sigma^2)$.

$\quad\quad t = \dfrac{1}{s/\sqrt{SS_{xx}}} = \dfrac{-0.3333}{\sqrt{0.1684}/\sqrt{8.25}} = -7.38$

Rejection Region: $t > -t_{0.05} = -1.860$ (degrees of freedom $= 10 - 2 = 8$)

Conclusion: Reject H_0 at $\alpha = 0.05$; i.e., the mean temperature decreases significantly as distance from the lake increases at $\alpha = 0.05$.

9.21

a.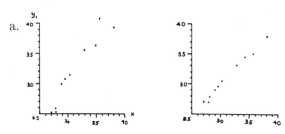

b. Both scattergrams indicate a possible linear relationship.

c. $\sum x_i = 312.8$ $\sum x_i^2 = 9,911.42$ $n = 10$

$\sum y_{1i} = 320.2$ $\sum y_{1i}^2 = 10,543.68$ $\sum x_i y_{1i} = 10,201.41$

$\sum y_{2i} = 311.2$ $\sum y_{2i}^2 = 9,809.52$ $\sum x_i y_{2i} = 9,859.84$

$$SS_{xx} = \frac{9,911.42 - (312.8)^2}{10} = 127.036$$

$$SS_{y_1 y_1} = \frac{10,543.68 - (320.2)^2}{10} = 290.876$$

$$SS_{xy_1} = \frac{10,201.41 - (312.8)(320.2)}{10} = 185.554$$

$$SS_{y_2 y_2} = \frac{9,809.52 - (311.2)^2}{10} = 124.976$$

$$SS_{xy_2} = \frac{9,859.84 - (312.8)(311.2)}{10} = 125.504$$

For x and y_1, $r_1 = \dfrac{185.554}{\sqrt{(127.036)(290.876)}} = 0.965$.

For x and y_2, $r_2 = \dfrac{125.504}{\sqrt{(127.036)(124.976)}} = 0.996$.

The r-values indicate the strength and direction of a linear relationship between the static and in-motion readings. The greater the r-value, the less variability will be displayed by the data about the least-squares line and hence the more precise the readings will be.

d. If the r-value is 1, then there is an exact linear relationship between y_2 and x; i.e., $y_2 = a + bx$. However, unless $a = 0$ and $b = 1$, the readings could still disagree.

9.23

a. $\sum x_{2i} = 10,166 \qquad \sum x_{2i}^2 = 37,987,830 \qquad \sum x_{2i}y_i = 4,586,410$

$$SS_{x_2x_2} = 37,987,830 - \frac{(10,166)^2}{10} = 27,653,074.4$$

$$SS_{x_2y} = 45,986,410 = \frac{(10,166)(10,595)}{10} = 35,215,533$$

$$\hat{\beta}_1 = \frac{SS_{x_2y}}{SS_{x_2x_2}} = \frac{35,215,533}{27,653,074.4} = 1.2735$$

$$\hat{\beta}_0 = \bar{y} - \hat{\beta}_1\bar{x} = \frac{10,595}{10} - (1.2735)\left(\frac{10,166}{10}\right) = -235.1159$$

Least-squares line: $\hat{y} = -235.1159 + 1.2735x$

b. Hypotheses: $H_0 : \beta_2 = 0 \qquad H_a : \beta_2 > 0$

Test Statistics: Assume that the errors are independent, normal $(0, \quad \sigma^2)$.

$$s^2 = \frac{SSE}{n-2} = \frac{45,917,560.5 - \dfrac{(35,215,533)^2}{27,653,074}}{8} = \frac{1,071,417.4}{8} = 133,927.026$$

$$t = \frac{1.2735}{\sqrt{133,927.026}/\sqrt{27,653,074.4}} = 18.30$$

Rejection Region: $t > t_{0.05} = 1.860 \qquad$ (degrees of freedom $= 8$)

Conclusion: Reject H_0 at $\alpha = 0.05$; i.e., there is sufficient evidence to conclude that the mean number of arrests increases as the number of law enforcement employees increases at $\alpha = 0.05$.

c. $r^2 = 1 - \dfrac{SSE}{SS_{yy}} = 1 - \dfrac{1,071,417.4}{45,917,560.5} = 0.9767$

About 97.7% of the variability in the number of drug arrests is accounted for by a linear relationship with the number of law enforcement employees at $\alpha = 0.05$.

9.25

a. $\hat{y} \pm t_{0.025}\sqrt{s^2\left(\dfrac{1}{n} + \dfrac{(x-\bar{x})^2}{SS_{xx}}\right)}$ (degrees of freedom $= 6 - 2 = 4$)

$= \left(\dfrac{6}{7}\right)(2) \pm 2.776\sqrt{(0.2857)\left(\dfrac{1}{6} + \dfrac{(2-3.5)^2}{17.5}\right)} = (0.9080, \quad 2.5205)$

b. $\hat{y} \pm t_{0.025}\sqrt{s^2\left(1 + \dfrac{1}{n} + \dfrac{(x-\bar{x})^2}{SS_{xx}}\right)}$ (degrees of freedom $= 4$)

$= \left(\dfrac{6}{7}\right)(2) \pm \sqrt{(0.2857)\left(1 + \dfrac{1}{6} + \dfrac{(2-3.5)^2}{17.5}\right)} = (0.02560, \quad 3.4030)$

9.27 $\hat{y} \pm t_{0.025}\sqrt{s^2\left(1 + \dfrac{1}{n} + \dfrac{(x-\bar{x})^2}{SS_{xx}}\right)}$ (degrees of freedom $= 6 - 2 = 4$)

$= -1.43 + 0.1947(16) \pm 2.776\sqrt{(0.009867)\left(1 + \dfrac{1}{6} + \dfrac{(16-13.1667)^2}{18.8333}\right)}$

$= (1.3370, \quad 2.0031)$

Industry sales figures in Exercise 9.4 range from 10 to 15 (millions of dollars). But we have no information about the behavior of the relationship between industry and company sales outside this range. Since the industry sales figure used here, 16, is outside the range 10 to 15, it is unadvisable to use the least-squares estimates.

9.29 $\hat{y} \pm t_{0.025}\sqrt{s^2\left(1 + \dfrac{1}{n} + \dfrac{(x-\bar{x})^2}{SS_{xx}}\right)}$ (degrees of freedom $= 6$)

$= 480.2043 - 3.7511(75) \pm 2.447\sqrt{19.1457\left(1 + \dfrac{1}{8} + \dfrac{(75-69.5)^2}{438}\right)}$

$= (187.1688, \quad 210.5687)$

Assume that the errors are independent, normal $(0, \quad \sigma^2)$.

9.31 $\hat{y} \pm t_{0.025}\sqrt{s^2 \left(1 + \dfrac{1}{n} + \dfrac{(x - \bar{x})^2}{SS_{xx}}\right)}$ (degrees of freedom $= 10 - 2 = 8$)

$$= -0.7110 + 0.6553(21) \pm 2.306\sqrt{10.9732 \left(1 + \dfrac{1}{10} + \dfrac{(21 - 15.504)^2}{2360.2388}\right)}$$

$$= (4.9911, 21.1074)$$

9.33 $\sum x_i = 21.65$ $\sum x_i^2 = 68.8357$ $\sum x_i y_i = 390.688$

   $\sum y_i = 124.7$ $\sum y_i^2 = 2235.75$ $n = 7$

   $SS_{xx} = 68.8357 - \dfrac{(21.65)^2}{7} = 1.8753$

   $SS_{yy} = 2235.75 - \dfrac{(124.7)^2}{7} = 14.3086$

   $SS_{xy} = 390.688 - \dfrac{(21.65)(124.7)}{7} = 5.0087$

   a. $\hat{\beta}_1 = \dfrac{SS_{xy}}{SS_{xx}} = \dfrac{5.0087}{1.8753} = 2.6708$

   $\hat{\beta}_0 = \bar{y} - \hat{\beta}_1 \bar{x} = \dfrac{124.7}{7} - (2.6708)\left(\dfrac{21.65}{7}\right) = 9.5538$

   Least-squares line: $\hat{y} = 9.5538 + 2.6708x$

b.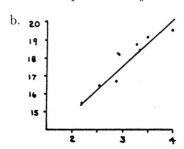

   Hypotheses: $H_0 : \beta_1 = 0$ $H_a : \beta_1 \neq 0$

   Test Statistics: Assume that errors are independent, normal $(0, \quad \sigma^2)$

   $s^2 = \dfrac{14.3086 - \dfrac{(5.0087)^2}{1.8753}}{5} = 0.1862$

   $t = \dfrac{2.6708}{\sqrt{0.1862}/\sqrt{1.8753}} = 8.48$

Rejection Region: $|t| > t_{0.025} = 2.571$ (degrees of freedom $= 7 - 2 = 5$)

Conclusion: Reject H_0 at $\alpha = 0.05$; i.e., there is a significant linear relationship between GPA in major courses and starting salary.

c. $\hat{y} \pm t_{0.025}\sqrt{s^2\left(1 + \dfrac{1}{n} + \dfrac{(x - \bar{x})^2}{SS_{xx}}\right)}$ (degrees of freedom $= 5$)

$= 9.5538 + 2.6708(3.2) \pm 2.571\sqrt{0.1862\left(1 + \dfrac{1}{7} + \dfrac{(3.2 - 3.0929)^2}{1.8753}\right)}$

$= (16.9112, \ 19.2897)$

d. $\hat{y} \pm t_{0.025}\sqrt{s^2\left(\dfrac{1}{n} + \dfrac{(x - \bar{x})^2}{SS_{xx}}\right)}$ (degrees of freedom $= 5$)

$= 9.5538 + 2.6708(3.0) \pm 2.571\sqrt{0.1862\left(\dfrac{1}{7} + \dfrac{(3 - 3.0929)^2}{1.8753}\right)}$

$= (17.1402, \ 17.9923)$

9.35 $\quad \sum x_i = 527.8 \qquad \sum x_i^2 = 39,904.8 \qquad \sum x_i y_i = 42,392.1$

$\sum y_i = 560.4 \qquad \sum y_i^2 = 45,072.92 \qquad n = 7$

$SS_{xx} = 39,904.8 - \dfrac{(527.8)^2}{7} = 108.68$

$SS_{yy} = 45,072.92 - \dfrac{(560)^2}{7} = 208.8971$

$SS_{xy} = 42,392.1 - \dfrac{(527.8)(560.4)}{7} = 137.94$

a. $r = \dfrac{137.94}{\sqrt{(108.68)(208.8971)}} = 0.9155$

b. $r^2 = (0.9155)^2 = 0.8381$ and represents the proportion of variability accounted for by a linear relationship.

c. To test if r is significantly different from zero, we may equivalently test whether or not $\hat{\beta}_1$ is significantly different from zero.

Hypotheses: $H_0 : \beta_1 = 0$ $\qquad\qquad$ $H_a : \beta_1 \neq 0$

Test Statistics: Assume that errors are independent, normal $(0, \quad \sigma^2)$.

$$s^2 = \frac{\left(208.8971 - \dfrac{(137.94)^2}{108.68}\right)}{5} = 6.7639$$

$$\hat{\beta}_1 = \frac{SS_{xy}}{SS_{xx}} = \frac{137.94}{109.8971} = 1.2692$$

$$t = \frac{1.2692}{\sqrt{6.7639}/\sqrt{108.68}} = 5.09$$

Rejection Region: $|t| > t_{0.025} = 2.571$ $\qquad$ (degrees of freedom $= 7 - 2 = 5$)

Conclusion: Reject H_0 at $\alpha = 0.05$; i.e., there is sufficient evidence at $\alpha = 0.05$ to indicate that the correlation differs from zero.

9.37 $\qquad \sum x_i = 124 \qquad\qquad \sum x_i^2 = 2,330 \qquad\qquad \sum x_i y_i = 2,820$

$\qquad\qquad \sum y_i = 150 \qquad\qquad \sum y_i^2 = 3,550 \qquad\qquad n = 7$

$$SS_{xx} = 2,330 - \frac{(124)^2}{7} = 133.4286$$

$$SS_{yy} = 3,550 - \frac{(150)^2}{7} = 335.7143$$

$$SS_{xy} = 2,820 - \frac{(124)(150)}{7} = 162.8571$$

$$\hat{\beta}_1 = \frac{SS_{xy}}{SS_{xx}} = \frac{162.8571}{133.4286} = 1.2206$$

$$\hat{\beta}_0 = \bar{y} - \hat{\beta}_1 \bar{x} = \left(\frac{150}{7}\right) - (1.2206)\left(\frac{124}{7}\right) = -0.1927$$

Least-squares line: $\hat{y} = -0.1927 + 1.2206x$

$$r^2 = 1 - \frac{SSE}{SS_{yy}} = 1 - \frac{\left(335.7143 - \dfrac{(162.8571)^2}{133.4286}\right)}{335.7143} = 0.5921$$

Thus, about 59% of the variability in the number of items produced is accounted for by a linear relationship between the total variable cost and total output.

9.39 $\sum x_i = 344$ $\sum x_i^2 = 8,716$ $\sum x_i y_i = 1,202,056$

$\sum y_i = 38,225$ $\sum y_i^2 = 199,395,657$ $n = 22$

$$SS_{xx} = 8,716 - \frac{(344)^2}{22} = 3,337.0909$$

$$SS_{yy} = 199,395,657 - \frac{(38,225)^2}{22} = 132,979,720$$

$$SS_{xy} = 1,202,056 - \frac{(344)(38,225)}{22} = 604,356$$

a. $\hat{\beta}_1 = \dfrac{SS_{xy}}{SS_{xx}} = \dfrac{604,356}{3,337.0909} = 181.1026$

$$\hat{\beta}_0 = \bar{y} - \hat{\beta}_1 \bar{x} = \frac{38,225}{22} - (181.1026)\left(\frac{344}{22}\right) = -1,094.2869$$

Least-squares line: $\hat{y} = -1,094.2869 + 181.1026x$

b. $s^2 = \dfrac{132,979,720 - \dfrac{(604,356)^2}{333.70909}}{20} = 1,176,462.37$

$\hat{\beta}_1 \pm t_{0.025} s / \sqrt{SS_{xx}}$ (degrees of freedom $= 22 - 2 = 20$)

$$= 181.1026 \pm 2.086 \frac{\sqrt{1,176,462.37}}{\sqrt{3,337.0909}} = (141.9357, 220.2695)$$

Since this interval does not include zero, we may conclude at a 95% confidence level that β_1 is significantly different from zero at $\alpha = 0.05$.

9.41 $\sum x_i = -2682.8$ $\sum x_i^2 = 720039.301$ $\sum x_i y_i = -1847.0073$

$\sum y_i = 6.826$ $\sum y_i^2 = 5.6326$ $n = 10$

$$SS_{xx} = 720039.301 - \frac{(-2682.8)^2}{10} = 297.717$$

$$SS_{yy} = 5.6326 - \frac{(6.826)^2}{10} = 0.9731$$

$$SS_{xy} = -1847.0073 - \frac{(-2682.8)(6.826)}{10} = -15.7280$$

a. $\hat{\beta}_1 = SS_{xy}/SS_{xx} = \dfrac{-15.7280}{297.717} = -0.05283$

$\hat{\beta}_0 = \bar{y} - \hat{\beta}_1\bar{x} = \dfrac{6.826}{10} - (-0.05283)\left(\dfrac{-2682.8}{10}\right) = -13.4904$

Least-squares line: $\hat{y} = -13.4904 - 0.05283x$

b. Hypotheses: $H_0 : \hat{\beta}_1 = 0$ $\qquad H_a : \beta_1 < 0$

Test Statistics: Assume that errors are independent, normal $(0, \quad \sigma^2)$.

$$s^2 = \frac{SSE}{(n-2)} = \frac{0.973 - \dfrac{(-15.7280)^2}{297.717}}{8} = 0.1423/8 = 0.01778$$

$$t = \frac{-0.05283}{\sqrt{0.01778}/\sqrt{297.717}} = -6.84$$

Rejection Region: $|t| > t_{0.01} = 2.896$ $\qquad$ (degrees of freedom $= 10 - 2 = 8$)

Conclusion: Reject H_0 at $\alpha = 0.01$; i.e., there is a significant linear relation between temperature and the proportion of impurity passing through the helium at $\alpha = 0.01$.

c. $r^2 = 1 - \dfrac{SSE}{SS_{yy}} = 1 - \dfrac{0.1423}{0.9731} = 0.8538$

Thus about 85% of the variability in the proportions of impurity passing through helium is accounted for by the linear relationship between temperature and the proportion of impurity passing through helium.

d. $\hat{y} \pm t_{0.025}\sqrt{s^2\left(1 + \dfrac{1}{n} + \dfrac{(x - \bar{x})^2}{S_{xx}}\right)} = -13.4904 - 0.05283(-273)$

$\pm\ 2.306\sqrt{0.01778\left(1 + \dfrac{1}{10} + \dfrac{(-273 - (-268.28))^2}{297.7171}\right)} = (0.5987,\ 1.2652)$

Since y is bounded above by 1, the interval becomes $(0.5987, 1)$.

9.43

  a. Hypotheses: $H_0 : \beta_1 = 0$ $\qquad$ $H_a : \beta_1 \neq 0$

  Rejection Region: $|t| > t_{0.025} = 2.576$ $\qquad$ (degrees of freedom $= 502(\infty)$)

Firm	Test Statistic	Conclusion at $\alpha = 0.05$
Conoco	t = 21.93	Model is adequate.
DuPont	t = 18.76	Model is adequate.
Mobil	t = 16.21	Model is adequate.
Seagram	t = 6.05	Model is adequate.

  b. The increase is given by $\Delta y = \hat{y}_{\text{new}} - \hat{y}_{\text{old}} = \hat{\beta}_0 + \hat{\beta}_1 x_{\text{new}} - (\hat{\beta}_0 + \hat{\beta}_1 x_{\text{old}})$
  $= \hat{\beta}_1(x_{\text{new}} - x_{\text{old}}) = \hat{\beta}_1(\Delta x)$. Seagram's mean rate of return would increase by $(0.76)(0.10) = 0.076$.

  c. Conoco's stocks appear to be more responsive since the slope of its least-squares regression line is greater; i.e., $1.40 > 0.76$.

9.45 $\quad \sum x_i = 230 \qquad\qquad \sum x_i^2 = 12,150 \qquad\qquad \sum x_i y_i = 9,850$

$\sum y_i = 215 \qquad\qquad \sum y_i^2 = 12,781 \qquad\qquad n = 5$

$SS_{xx} = 12,150 - \dfrac{(230)^2}{5} = 1,570$

$SS_{yy} = 12,781 - \dfrac{(215)^2}{5} = 3,536$

$SS_{xy} = 9,850 - \dfrac{(230)(215)}{5} = -40$

a. $\hat{\beta}_1 = \dfrac{SS_{xy}}{SS_{xx}} = \dfrac{-40}{1,570} = -0.02548$

$\hat{\beta}_0 = \bar{y} - \hat{\beta}_1 \bar{x} = \dfrac{215}{5} - (-0.02548)\left(\dfrac{230}{5}\right) = 44.1720$

Least-squares line: $\hat{y} = 44.1720 - 0.02548x$

b.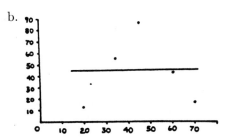

c. Hypotheses: $H_0 : \beta_1 = 0 \qquad H_a : \beta_1 \neq 0$

Test Statistics: Assume that errors are independent, normal $(0, \ \sigma^2)$.

$$s^2 = \dfrac{SSE}{(n-2)} = \dfrac{3536 - \dfrac{(-40)^2}{1570}}{3} = \dfrac{3534.9807}{3} = 1178.3270$$

$$t = \dfrac{-0.02548}{\sqrt{1178.3270}/\sqrt{1570}} = -0.03$$

Rejection Region: $|t| > t_{0.025} = 3.182 \qquad$ (degrees of freedom $= 3$)

Conclusion: Fail to reject H_0 at $\alpha = 0.05$; i.e., there is not a significant linear relationship between tire price and number sold at $\alpha = 0.05$.

d. The nonrejection of H_0 in part (c) simply indicates that no significant *linear* relationship exists; however, a curvilinear relationship may still be present.

e. $r^2 = 1 - \dfrac{SS}{SS_{yy}} = 1 - \dfrac{3,534.98907}{3,536} = 0.0002882$ gives the proportion of variability in the number sold that is accounted for by a linear relationship between tire price and the number sold.

9.47 $\quad \sum x_i = 741 \qquad\qquad \sum x_i^2 = 68,789 \qquad\qquad \sum x_i y_i = 10,016.3$

$\sum y_i = 108 \qquad\qquad \sum y_i^2 = 1,459.34 \qquad\qquad n = 8$

$$SS_{xx} = 68,789 - \frac{(741)^2}{8} = 153.875$$

$$SS_{yy} = 1,459.34 - \frac{(108)^2}{8} = 1.340$$

$$SS_{xy} = 10,016.3 - \frac{(741)(108)}{8} = 12.8$$

a. $r = \dfrac{12.8}{\sqrt{(153.875)(1.340)}} = 0.89$

$r^2 = (0.89)^2 = 0.79$

b. We need to test to determine if r is significantly different from zero. Equivalently, we test to determine whether or not the slope is significantly different from zero.

Hypotheses: $H_0 : \beta_1 = 0$ $\qquad$ $H_a : \beta_1 \neq 0$

Test Statistic: Assume that the curves are independent, normal $(0, \quad \sigma^2)$.

$$s^2 = \frac{1.340 - \dfrac{(12.8)^2}{153.875}}{6} = 0.04587$$

$$\hat{\beta}_1 = \frac{SS_{xy}}{SS_{xx}} = \frac{12.8}{153.875} = 0.08318$$

$$t = \frac{0.08318}{\sqrt{0.04587}/\sqrt{153.875}} = 4.82$$

P-value: Let $p = P(|t| > 4.82)$; then $p/2 < 0.005$ so that $p < 0.01$.

Conclusion: Since the p-value is small (say, less than 0.05), we reject H_0 and conclude that the correlation significantly differs from zero.

"MINITAB" OUTPUT WILL BE GIVEN FOR

A SUGGESTED SOLUTION TO THE PROBLEMS

9.49   The regression equation is $\ln(y) = -64.34 + 0.04x$ where y = cumulative number

of stamps issued and x = year.

Predictor	Coef	Stdev	t-ratio	p
Constant	−64.3402	9.8979	−6.500	0.0001
x	0.0366	0.0052	7.085	0.0001

$s = 0.8633$ R-sq = 79.43%   R-sq(adj) = 77.85%

Analysis of Variance

Source	DF	SS	MS	F	P
Regression	1	37.4118	37.4118	50.198	0.0001
Error	13	9.6887	0.7453		
		47.1005			

Unusual Observations

Obs	x	ln(Y)	Fit	Stdev.Fit	Residual	St.Residual
1	2	0.6931	3.2101	0.424	−2.5169	−3.348R

R denotes an obs. with a large st. resid.

Residual plot is in the Appendix.

9.51   The regression equation is matter = $910 - 0.455$ year.

Predictor	Coef	Stdev	t-ratio	p
Constant	909.83	87.18	10.44	0.000
year	−0.45479	0.04418	−10.29	0.000

$s = 2.187$ R-sq = 86.2% R-sq(adj) = 85.4%

Analysis of Variance

Source	DF	SS	MS	F	p
Regression	1	506.85	506.85	105.96	0.000
Error	17	81.32	4.78		
Total	18	588.17			

9.53 Current dollars

For men the regression equation is

Salary = −1, 975, 417 + 1, 007.5879 (year)

Predictor	Coef	Stdev	t-ratio	p
Constant	−1975417	64811.1538	30.852	0.0001
year	1007.5879	32.6586	30.852	0.0001

Analysis of Variance

Source	DF	SS	MS	F	p
Regression	1	83756750	83756750	951.852	0.0001
Error	8	703947.6	87993.4		
Total	9	84460697.6			

For women the regression equation is

salary = −1,694,230 + 861.6121 (year)

Predictor	Coef	Stdev	t-ratio	p
Constant	−1694230	42532.2537	−39.834	0.0001
year	861.6121	21.4322	40.202	0.001

Analysis of Variance

Source	DF	SS	MS	F	P
Regression	1	61245974	61245974	1616.181	0.0001
Error	8	303164.0	37895.5		
Total	9	61549138			

Constant dollars

For men the regression equation is

salary = −89407 + 59.4606 (year)

Predictor	Coef	Stdev	t-ratio	p
Constant	−89407	101200.3586	−0.879	0.4050
year	59.4606	51.2473	1.160	0.2794

Analysis of Variance

Source	DF	SS	MS	F	p
Regression	1	291684.00	291684.00	1.346	0.2794
Error	8	1733348.10	216668.51		
Total	9	2025032.10			

For women the regression equation is

salary $= -559{,}303 + 291.1273$ (year)

Predictor	Coef	Stdev	t-ratio	p
constant	-559303	48089.6629	-11.630	0.0001
year	291.1273	24.2326	12.014	0.0001

Analysis of Variance

Source	DF	SS	MS	F	p
Regression	1	6992294.8	6992294.8	144.333	0.0001
Error	8	387564.8	48445.6		
Total	9	7379859.6			

9.55 The regression equation is content $= -0.3213 + 12.3081$ absorbance.

Predictor	Coef	Stdev	t-ratio	p
Constant	-0.3213	0.0043	-74.63	0.0000
absorbance	12.3081	0.0150	822.38	0.0000

Analysis of Variance

Source	DF	SS	MS	F	p
Regression	1	810.9320	810.9320	676300.76	0.0000
Error	99	0.1187	0.1187		
Total	100	811.05070			

If all the data were used in finding the linear model, then the correlation would increase.

Chapter 10.

10.1

    a. A value of $R^2 = 0.89$ means that 89% of the variation in the y's is accounted for by the model. This suggests that the model may provide a good fit to the data.

    b. Hypotheses: $H_0 : \beta_1 = \cdots = \beta_5 = 0$

      H_a: At least one β_i is nonzero, $i = 1, ..., 5$

      Test Statistic: $F = \dfrac{R^2/k}{(1 - R^2)/(n - (k+1))} = \dfrac{0.89/5}{(0.11)/(30 - 6)} = 38.84$

      Rejection Region: $F > F_{0.05}(k, \quad n - (k+1)) = F_{0.05}(5, \quad 24) = 2.62$

      Conclusion: Reject H_0 at $\alpha = 0.05$; i.e., the fit of the model is adequate at $\alpha = 0.05$ and hence the model is of some use in predicting y.

10.3   Hypotheses: $H_0 : \beta_1 = \cdots = \beta_{18} = 0$

      H_a: At least one β_i is nonzero, $i = 1, ..., 18$

      Test Statistic: $F = \dfrac{0.95/18}{(1 - 0.95)/(20 - 19)} = 1.06$

      Rejection Region: $F > F_{0.05}(18, 1) \approx 247.16$ by interpolating from F table.

      Conclusion: Fail to reject H_0 at $\alpha = 0.05$; i.e., none of the β parameters are significantly different from zero, so the model is not adequate at $\alpha = 0.05$.

10.5

    a. $\hat{y} = 20.09111 - 0.6705x + 0.009535x^2$

    b.

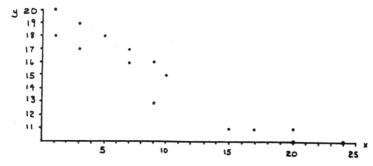

c. Hypotheses: $H_0 : \beta_2 = 0$ $\qquad$ $H_a : \beta_2 \neq 0$

Test Statistic: $t = \dfrac{\hat{\beta}_2}{s_{\hat{\beta}_2}} = \dfrac{0.009535}{0.006326} = 1.51$

Rejection Region: $|t| > t_{0.01} = 3.055$ $\qquad$ (degrees of freedom = 12)

Conclusion: Fail to reject H_0 at $\alpha = 0.01$, i.e., the quadratic term does not make a significant contribution to the model.

d. $\sum x_i = 151$ $\qquad$ $\sum x_i^2 = 2,295$ $\qquad$ $\sum x_i y_i = 1,890$

$\sum y_i = 222$ $\qquad$ $\sum y_i^2 = 3,456$ $\qquad$ $n = 15$

$SS_{xx} = 2,295 - \dfrac{(151)^2}{15} = 774.9333$

$SS_{yy} = 3,456 - \dfrac{(222)^2}{15} = 170.4$

$SS_{xy} = 1,890 - \dfrac{(151)(222)}{15} = -344.8$

$\hat{\beta}_1 = \dfrac{SS_{xy}}{SS_{xx}} = \dfrac{-344.8}{774.9333} = -0.4449$

$\hat{\beta}_0 = \bar{y} - \hat{\beta}_1 \bar{x} = \dfrac{222}{15} - (-0.4444)\left(\dfrac{151}{15}\right) = 19.2791$

Reduced fitted model: $\hat{y} = 19.2791 - 0.4449x$

(To test the utility of the model, we compute $s^2 = 1.3065$, which yields a $t = \hat{\beta}_1 / \left(s/\sqrt{SS_{xx}}\right) = -10.84$ for testing $H_0 : \beta_1 = 0$ vs. $H_a : \beta_1 \neq 0$. This is highly significant, so the model is adequate.)

e. $\hat{\beta}_1 \pm \dfrac{t_{0.05} s}{\sqrt{SS_{xx}}} = -0.4449 \pm \dfrac{1.771\sqrt{1.3065}}{\sqrt{774.9333}} = (-0.5177, -0.3722)$

(degrees of freedom = 13)

In the context of this problem, β_1 is the decrease in time required to complete the task per additional month of experience.

10.7

  a. Hypotheses: $H_0 : \beta_2 = 0 \qquad H_a : \beta_2 \neq 0$

  Test Statistic: $t = \dfrac{\hat{\beta}_2}{s_{\hat{\beta}_2}} = \dfrac{0.55}{0.181} = 3.04$

  P-value: Let $p = P(|t| > 3.04)$; then $0.005 < p/2 = P(t > 3.04) < 0.01$ so that $0.01 < p < 0.02$ (degrees of freedom $= 8$).

  Conclusion: Since the P-value is small (say, less than 0.05), we reject H_0 and conclude that the mean Christmas sales are related to August sales in the proposed model.

10.9 If there are not interaction terms in the model and m independent variables, then the number of degrees of freedom for error is $n - (m + 1)$. If there are m independent variables and all possible interactions are included in the full model, then (for $n > 2^m$) there will be $n - 2^m$ degrees of freedom for error. The number of degrees of freedom for error is most easily related to k, the number of parameters (excluding β_0) giving $n - (k + 1)$ degrees of freedom for error.

10.11

  a. $Y = \beta_0 + \beta_1 x_1 + \beta_2 x_1^2 + \epsilon$

  b. Hypotheses: $H_0 : \beta_1 = \beta_2 = 0 \qquad H_a : \beta_1$ or β_2 is nonzero

  Test Statistic: $F = 44.50$

  Rejection Region: $F > F_{0.05}(2, \quad 27) = 3.35$

  Conclusion: Reject H_0 at $\alpha = 0.05$; i.e., the model is adequate at $\alpha = 0.05$.

  c. The P-value of 0.0001 indicates that if this experiment were repeated, only about once in 10,000 trials would an F-value as high as 44.5 be observed by chance when, in fact, $\beta_1 = \beta_2 = 0$.

d. Hypotheses: $H_0 : \beta_2 = 0 \qquad H_a : \beta_2 \neq 0$

Test Statistic: $t = -4.68$

Rejection Region: $|t| > t_{0.025} = 2.052$ $\qquad$ (degrees of freedom $= 27$)

Conclusion: We reject H_0 at $\alpha = 0.05$ and conclude that the second-order term is significant.

e. The P-value of 0.0001 indicates that if this experiment were repeated, only about once in 10,000 trials would a t-value at least 4.68 units from zero be observed by chance when, in fact, $\beta_2 = 0$.

10.13

a. We would fit the complete model

$$Y = \beta_0 + \beta_1 x_1 + \beta_2 x_2 + \beta_3 x_3 + \beta_4 x_4 + \epsilon$$

where $k = 4$ in order to determine SSE_2. Then we would fit the reduced model $Y = \beta_0 + \beta_1 x_3 + \beta_2 x_4 + \epsilon$ where $g = 2$ in order to determine SSE_1. These are the relevant quantities for the F-statistic.

b. $\nu_1 = k - g = 4 - 2 = 2 \qquad\qquad \nu_2 = n - (k+1) = 25 - (4+1) = 20$

10.15

a. $H_0 : \beta_1 = \cdots = \beta_5 = 0$

H_a: At least one of $\beta_1, \ldots, \beta_5$ is nonzero

b. $H_0 : \beta_3 = \beta_4 = \beta_5 = 0$

H_a: At least one of β_1, β_4, or β_5 is nonzero

c. Test Statistic: $F = \dfrac{0.729/5}{(1 - 0.729)/(40 - 6)} = 18.29$

Rejection Region: $F > F_{0.05}(5, 34) \approx 2.50$

Conclusion: Reject H_0 at $\alpha = 0.05$; i.e., the complete model is useful for prediction since at least one of the β's is significantly different from zero at $\alpha = 0.05$.

d. Test Statistic: $F = \dfrac{(3,197.16 - 1,830.44)/(5-2)}{1,830.44/(40-(5+1))} = 8.46$

   Rejection Region: $F > F_{0.05}(3, 34) \approx 2.89$

   Conclusion: Reject H_0 at $\alpha = 0.05$; i.e., the second order model is significant at $\alpha = 0.05$.

10.17 Hypotheses: $H_0 : \beta_1 = \beta_3 = 0$ $\qquad$ $H_a : \beta_2 \neq 0$ or $\beta_3 \neq 0$

   Test Statistic: $F = \dfrac{(795.23 - 783.90)/2}{783.90/(200-(4+1))} = 1.41$

   Rejection Region: $F > F_{0.05}(2, 195) \approx 3.06$

   Conclusion: Fail to reject H_0 at $\alpha = 0.05$; i.e., there is insufficient evidence that the mean faculty salary is dependent on sex at $\alpha = 0.05$.

10.19

a. $\hat{y} = 0.04565 + 0.000785x_1 + 0.23737x_2 - 0.0000381x_1x_2$

b. $SSE = 2.7152, s^2 = MSE = 0.1697$

c. The least-squares technique chooses estimates for β_0, β_1, β_2, β_3 so as to minimize

   $SSE = \sum_i (y_i - \hat{y}_i)^2$.

10.21 From the SAS printout, a 95% confidence interval for the mean cost of computer jobs that require 42 seconds of CPU time and print 2,000 lines is ($7.32, $9.45).

10.23

a. $\hat{y} = 0.6013 + 0.5953x_1 - 3.7254x_2 - 16.2320x_3 + 0.2349x_1x_2 + 0.3081x_1x_3$

b. The value of $R^2 = 0.9281$ indicates that about 92.8% of the variability in the Y scores is accounted for by the model.

   Hypotheses: $H_0 : \beta_1 = \cdots = \beta_5$

   H_a: At least one of $\beta_1, \ldots, \beta_5$ is nonzero

   Test Statistic: F = 139.42

   P-value: 0.0001

Conclusion: Since the P-value of 0.0001 is extremely small, we reject H_0 and we have sufficient evidence that the model is useful for predicting achievement test scores.

c.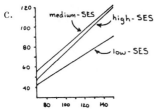

d. Hypotheses: $H_0 : \beta_4 = \beta_5 = 0$

$H_a : \beta_4 \neq 0$ or $\beta_5 \neq 0$

Test Statistic: $F = \dfrac{(1216.0189 - 969.4831)/(5 - 3)}{969.4831/(60 - (5 + 1))} = 6.87$

Rejection Region: $F > F_{0.05}(2, 54) \approx 3.20$

Conclusion: Reject H_0 at $\alpha = 0.05$; i.e., the mean increase in achievement test scores per unit increase in IQ significantly differs for the three levels of SES at $\alpha = 0.05$.

10.25   Hypotheses: $H_0 : \beta_2 = 0$ $\qquad\qquad$ $H_a : \beta_2 < 0$

Test Statistic: $t = \dfrac{-0.53}{0.48} = -1.10$

Rejection Region: $t < -t_{0.01} = -2.326$

(degrees of freedom $= 42 - 3 = 39(\infty)$)

Conclusion: Fail to reject H_0 at $\alpha = 0.01$; i.e., there is insufficient evidence to conclude that after allowing for the effect of initial assembly time, plant A had a lower mean assembly time than plant B.

10.27   We test the hypotheses from Exercise 10.26, part (b).

Test Statistic: $F = \dfrac{(259.34 - 226.12)/(4 - 2)}{226.12/(50 - (4 + 1))} = 3.31$

Rejection Region: $F > F_{0.05}(2, 45) \approx 3.21$

Conclusion: Reject H_0 at $\alpha = 0.05$; i.e., the mean delivery time significantly differs for mail and truck deliveries at $\alpha = 0.05$.

10.29

    a. Y = cost of material and labor

       x_1 = area

       x_2 = number of baths

$$x_3 = \begin{cases} 1 & \text{central air} \\ 0 & \text{no central air} \end{cases}$$

       First-order model: $Y = \beta_0 + \beta_1 x_1 + \beta_2 x_2 + \beta_3 x_3 + \epsilon$

    b. Second-order model: $Y = \beta_0 + \beta_1 x_1 + \beta_2 x_2 + \beta_3 x_3 + \beta_4 x_1 x_2 + \beta_5 x_1 x_3$

       $+ \beta_6 x_2 x_3 + \beta_7 x_1^2 + \beta_8 x_2^2 + \epsilon$

    c. $H_0 : \beta_4 = \cdots = \beta_8 = 0$

       H_a: At least one of $\beta_4, \ldots, \beta_8$ is nonzero

10.31   (Uses SAS output: see the Appendix.)

    a. $\hat{y} = 1.6794 + 0.4441 x_1 - 0.0793 x_2$

    b. Hypotheses: $H_0 : \beta_1 = \beta_2 = 0$ $H_a : \beta_1 \neq 0$ or $\beta_2 \neq 0$

       Test Statistic: $F = 150.62$

       P-value: 0.0001

       Conclusion: Since the P-value is extremely small, we reject H_0 and conclude that the model is appropriate.

    c. Since $R^2 = 0.9773$ is large, it tends to support the finding that the model is appropriate.

    d. $(-0.0793 \pm 1.895(0.005981)) = (-0.0907, -0.0680)$

       Since this interval does not include zero, we may conclude at a 90% confidence level that β_2 is not zero. Hence the service person's number of months of experience in preventive maintenance is useful in predicting time of preventive maintenance.

    e. $\hat{y} = 1.6794 + 0.4441(0) - 0.0793(6) = 1.2036$ (hours)

    f. Assuming that the service times are independent, the predicted mean time to service ten computers is 12.036 hours.

    g. (1.6430, 1.9785)

10.33   (Uses SAS output: see the Appendix.)

$$\hat{y} = -9.9168 + 0.1668x_1 + 0.1376x_2 - 0.001108x_1^2 - 0.0008433x_2^2 + 0.0002411x_1x_2$$

a. The value of $R^2 = 0.9365$ indicates that about 93.65% of the variability in the GPA data is accounted for by the model.

Hypotheses: $H_0 : \beta_1 = \cdots = \beta_5 = 0$

H_a: At least one of $\beta_1, \ldots, \beta_5$ is nonzero

Test Statistic: F = 100.41

P-value: $0.0001 < 0.05 = \alpha$

Conclusion: Since the P-value is extremely small, we reject H_0 and conclude that the model is useful in predicting mean freshman GPA values.

b.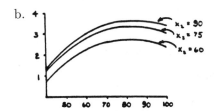

c. Hypotheses: $H_0 : \beta_5 = 0$ $H_a : \beta_5 \neq 0$

Test Statistic: t = 1.67

P-value: $0.1032 > 0.10 = \alpha$

Conclusion: Fail to reject H_0 at $\alpha = 0.10$; i.e., there is insufficient evidence to conclude that the interaction term is important for the prediction of GPA.

10.35   (Uses SAS output: see the Appendix.)

a.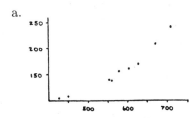

b. $\hat{y} = -93.1277 + 0.4446x$

c. $\hat{y} = 204.4603 - 0.6380x + 0.0009593x^2$

d. Hypotheses: $H_0 : \beta_2 = 0$ $H_a : \beta_2 \neq 0$

Test Statistic: t = 3.64

P-value: 0.0108

Conclusion: Since the P-value is small, we reject H_0 and conclude that the quadratic model is useful in describing the relationship between sulfur dioxide and output.

 e. $\hat{y} = 204.4603 - 0.6380(500) + 0.0009593(500)^2 = 125.2853$

10.37   Hypotheses: $H_0 : \beta_2 = 0$ $H_a : \beta_2 < 0$

    Test Statistic: t $= -6.60$

    Rejection Region: $t < -t_{0.05} = -1.717$ (degrees of freedom $= 22$)

    Conclusion: Reject H_0 at $\alpha = 0.04$; i.e., there is sufficient evidence to conclude that the rate of increase in output per unit increase of input decreases as the input increases.

10.39

 a. Hypotheses: $H_0 : \beta_1 = \beta_2 = 0$ $H_a : \beta_1 \neq 0$ or $\beta_2 \neq 0$

   Test Statistics: We fit the reduced model $Y = \beta_0 + \beta_3 x_2 + \epsilon$.

   Summary Statistics: $\sum x_2 = \sum x_2^2 = 10$, $\sum y = 461$, $\sum y^2 = 13151$, $\sum xy = 228$. Then for the reduced model $SSE = SS_{yy} - SS_{xy}^2/SS_{xx} = 2554.95 - (-2.5)^2/5 = 2523.70$.

$$F = \frac{(2523.7 - 128.586)/2}{128.586(20 - 4)} = 149.01$$

   Rejection Region: $F > F_{0.01}(2, \quad 16) > F_{0.05}(2, \quad 16) = 3.63$

 b. Conclusion: Reject H_0 at $\alpha = 0.05$; i.e., there is a significant quadratic relationship between age of machine and time for repairs.

10.41   Hypotheses: $H_0 : \beta_3 = \beta_4 = \beta_5 = 0$

    H_a: At least one of β_3, β_4, or β_5 is nonzero

    Test Statistic: $F = \dfrac{(370.7911 - 164.9185)/3}{164.9185/(30 - 6)} = 9.99$

    Rejection Region: $F > F_{0.05}(3, 24) = 3.01$

    Conclusion: Reject H_0 at $\alpha = 0.05$; i.e., the inclusion of the variable for speed limit contributes information for the prediction of number of highway deaths.

10.43   (Refer to SAS output in Appendix)

     a. Reaction = 1.1 speed

       Reaction = 1.1(55) = 60.5

     b. Braking = $-102.4952 + 4.8629$ speed

       Braking = $-102.4952 + 4.8629(55) = 164.9643$

     c. Total = $-102.4952 + 5.9629$ speed

       Total = $-102.4952 + 5.9629(55) = 225.4643$

10.45   Model for BaP: %CN = $-25.5902 + 68.8217$ mean Rf

       Model for BaA: %CN = $-20.6338 + 77.4706$ mean Rf

       Model for Phe: %CN = $-16.0972 + 86.0228$ mean Rf

Chapter 11.

11.1 Summary Statistics: $\sum y_i = 28.8$, $\sum y_i^2 = 106.32$,

$T_1 = 8.6$, $T_2 = 15.9$, $T_3 = 4.3$

a. ANOVA Table

Source	df	SS	MS	F
Treatments	2	11.0752	5.5376	3.1513
Error	7	12.3008	1.7573	
Total	9	23.3760		

b. Hypotheses: $H_0 : \mu_1 = \mu_2 = \mu_3$ H_a: At least two means differ

Test Statistic: F = 3.15

Rejection Region: $F > F_{0.05}(2, 7) = 4.74$

Conclusion: Fail to reject H_0 at $\alpha = 0.05$; i.e., the treatment means are not significantly different at $\alpha = 0.05$.

11.3 Summary Statistics: $\sum y_i = 1155$, $\sum y_i^2 = 103083$,

$T_1 = 244$, $T_2 = 269$, $T_5 = 381$, $T_7 = 261$

ANOVA Table

Source	df	SS	MS	F
Treatments	3	345.6089	115.2030	8.6342
Error	9	120.0833	13.3426	
Total	12	465.6922		

Hypotheses: $H_0 : \mu_1 = \mu_2 = \mu_3 = \mu_5 = \mu_7$ H_a: At least two means differ

Test Statistic: F = 8.63

Rejection Region: $F > F_{0.05}(3, \quad 9) = 3.86$

Conclusion: Reject H_0 at $\alpha = 0.05$; i.e., the mean percentage of kill significantly differs for the four rates of application of nematicide at $\alpha = 0.05$.

11.5 (a) and (b):

$$\text{Summary Statistics: } \bar{y} = \frac{42.960 + 145.222 + 92.249 + 73.224}{109} = 3.2445,$$

$$s_p^2 = \frac{2.0859 + 14.0186 + 7.9247 + 5.6812}{105} = 0.2830,$$

$$F = \frac{\left(15(2.864 - \bar{y})^2 + 41(3.542 - \bar{y})^2 + 29(3.181 - \bar{y})^2 + 24(3.051 - \bar{y})^2\right)}{3s_p^2}$$

$$= 8.0283$$

ANOVA Table

Source	df	SS	MS	F
Treatments	3	6.8160	2.2720	8.0283
Error	105	29.7150	0.2830	
Total	108	36.5310		

c. Hypotheses: $H_0 : \mu_1 = \mu_2 = \mu_3 = \mu_4$ H_a: At least two means differ

Test Statistic: F $= 8.0283$

Rejection Region: $F > F_{0.05}(3, \quad 105) \approx 2.70$

Conclusion: Reject H_0 at $\alpha = 0.05$; i.e., the mean scores are significantly different for the four academic ranks.

d. $2.864 - 3.051 \pm 1.96\sqrt{0.2830\left(\frac{1}{15} + \frac{1}{24}\right)} = (-0.5302, \quad 0.1562)$

e. $\dfrac{42.960 + 145.222}{15 + 41} - \dfrac{92.249 + 73.224}{29 + 24} \pm 1.96\sqrt{0.2830\left(\frac{1}{56} + \frac{1}{53}\right)}$

$= (0.0384, 0.4381)$

Since this interval does not contain zero, we may conclude at the 95% confidence level ($\alpha = 0.05$) that there is a significant difference between scores for tenured and nontenured faculty members.

11.7

a. ANOVA Table

Source	df	SS	MS	F
Company	1	3237.2	3237.2	19.6222
Error	98	16167.7	164.9765	
Total	99	19404.9		

b. Hypotheses: $H_0 : \mu_1 = \mu_2 \qquad H_a : \mu_1 \neq \mu_2$

Test Statistic: $F = 19.62$

Rejection Region: $F > F_{0.05}(1, 98) \approx 3.95$

Conclusion: Reject H_0 at $\alpha = 0.05$; i.e., there is a significant difference between the number of hours missed for the two companies.

11.9 Hypotheses: $H_0 : \mu_1 = \mu_2 = \mu_3 = \mu_4 \qquad H_a$: At least two means differ

Test Statistics: $\bar{y} = \dfrac{(8(80) + 8(81) + 8(86) + 8(90))}{32} = 84.25$

$s_p^2 = \dfrac{700}{(32 - 4)} = 25$

$F = \dfrac{\left(8(80 - 84.25)^2 + \cdots + 8(90 - 84.25)^2\right)}{(25)(3)} = 6.9067$

Rejection Region: $F > F_{0.05}(3, 28) = 2.95$

Conclusion: Reject H_0 at $\alpha = 0.05$; i.e., there are significant differences among the mean percentages of copper for the four castings.

11.11 Summary Statistics: $\sum y_i = 772, \sum y_i^2 = 30,550,$

$T_A = 174, T_B = 208, T_C = 231, T_D = 159,$

$\text{TSS} = 30,550 - \dfrac{(722)^2}{20} = 750.8,$

$\text{SST} = \dfrac{(174)^2}{5} + \cdots + \dfrac{(159)^2}{5} - \dfrac{(722)^2}{20} = 637.2$

ANOVA Table

Source	df	SS	MS	F
Treatments	3	637.2	212.4	29.9
Error	16	113.6	7.1	
Total	19	750.8		

Hypotheses: $H_0 : \mu_A = \mu_B = \mu_C = \mu_D$ H_a: At least two means differ

Test Statistic: $F = 29.9$

Rejection Region: $F > F_{0.05}(3, \quad 16) = 3.24$

Conclusion: Reject H_0 at $\alpha = 0.05$; i.e., there are significant differences among the mean tensile strengths for the four heat treatments.

11.13   Model: $Y = \beta_0 + \beta_1 x_1 + \beta_2 x_2 + \epsilon$ where

$$x_1 = \begin{cases} 1 & \text{sample 1} \\ 0 & \text{otherwise} \end{cases} \qquad x_2 = \begin{cases} 1 & \text{sample 2} \\ 0 & \text{otherwise} \end{cases}$$

Sample	x_1	x_2	E(Y)
1	1	0	$\beta_0 + \beta_1 = \mu_1$
2	0	1	$\beta_0 + \beta_2 = \mu_2$
3	0	0	$\beta_0 = \mu_3$

Hypotheses: Since $\beta_1 = \mu_1 - \mu_3$ and $\beta_2 = \mu_2 - \mu_3, H_0 : \mu_1 = \mu_2 = \mu_3$ is equivalent to $H_0 : \beta_1 = \beta_2 = 0$ with $H_a : \beta_1 \neq 0$ or $\beta_2 \neq 0$.

The ANOVA table, test statistic, rejection region, and conclusion are exactly as given in Exercise 11.1.

11.15   Model: $Y = \beta_0 + \beta_1 x_1 + \beta_2 x_2 + \beta_3 x_3 + \epsilon$ where

$$x_1 = \begin{cases} 1 & \text{thermometer 1} \\ 0 & \text{otherwise} \end{cases} \qquad x_2 = \begin{cases} 1 & \text{thermometer 2} \\ 0 & \text{otherwise} \end{cases}$$

Thermometer	x_1	x_2	E(Y)
1	1	0	$\beta_0 + \beta_1 = \mu_1$
2	0	1	$\beta_0 + \beta_2 = \mu_2$
3	0	0	$\beta_0 = \mu_3$

Hypotheses: Since $\beta_1 = \mu_1 - \mu_3$ and $\beta_2 = \mu_2 - \mu_3$,

$H_0 : \mu_1 = \mu_2 = \mu_3$ is equivalent to $H_0 : \beta_1 = \beta_2 = 0$ with $H_a : \beta_1 \neq 0$ or $\beta_2 \neq 0$.

The ANOVA table, test statistic, rejection region and conclusion are exactly as given in Exercise 11.10.

11.17

a. Hypotheses: $H_0 : \mu_1 = \mu_2 \qquad H_a : \mu_1 \neq \mu_2$

Test Statistic: $t = \dfrac{3.7 - 4.1}{\sqrt{1.2567 \left(\dfrac{1}{7} + \dfrac{1}{7} \right)}} = -0.67$

Rejection Region: $|t| > 1.645 \qquad$ (degrees of freedom $= 30$)

Conclusion: Fail to reject H_0 at $\alpha = 0.10$; i.e., there is not a significant difference between μ_1 and μ_2 at $\alpha = 0.10$.

b. $3.7 - 4.1 \pm 1.645 \sqrt{1.2567 \left(\dfrac{1}{7} + \dfrac{1}{7} \right)} = (-1.3857, \quad 0.5857)$

c. $3.7 \pm 1.645 \sqrt{\dfrac{1.2567}{7}} = (3.0030, 4.3970)$

11.19

a. $\dfrac{33.6}{5} - \dfrac{44.1}{5} \pm 2.179 \sqrt{0.8620 \left(\dfrac{1}{5} + \dfrac{1}{5} \right)} = (-3.3795, -0.8205)$

b. $t_{0.1/6} \approx 2.401 \qquad$ (degrees of freedom $= 12$)

i	j	$y_i - y_j \pm 2.401 \sqrt{0.8620 \left(\frac{1}{n_1} + \frac{1}{n_2} \right)}$
1	2	(5.7231, 7.7169)
1	3	(7.1231, 9.1169)
2	3	(7.8231, 9.8169)

11.21 Summary Statistics: $\sum y_i = 43$, $\sum y_i^2 = 197$, $T_1 = 12$, $T_2 = 22$, $T_3 = 9$,

$B_1 = 10$, $B_2 = 16$, $B_3 = 7$, $B_4 = 10$, $n = 12$,

$$TSS = 197 - \frac{(43)^2}{12} = 2.9167,$$

$$SST = \frac{1}{4}\left((12)^2 + (22)^2 + (9)^2\right) - \frac{(43)^2}{12} = 23.1667,$$

$$SSB = \frac{1}{3}\left((10)^2 + (16)^2 + (7)^2 + (10)^2\right) - \frac{(43)^2}{12} = 14.25$$

a. ANOVA Table

Source	df	SS	MS	F
Treatment	2	23.1667	11.5833	12.6333
Block	3	14.25	4.7500	5.1889
Error	6	5.5	0.9167	
Total	11	42.9167		

b. Hypotheses: $H_0 : \mu_1 = \mu_2 = \mu_3$ H_a: At least two means differ

Test Statistic: F = 12.63

Rejection Region: $F > F_{0.05}(2, 6) = 5.14$

Conclusion: Reject H_0 at $\alpha = 0.05$; i.e., there are significant differences among the treatment means.

c. Hypotheses: H_0: Block means are equal

H_a: At least two block means differ

Test Statistic: F = 5.19

Rejection Region: $F > F_{0.05}(3, 6) = 4.76$

Conclusion: Reject H_0 at $\alpha = 0.05$; i.e., we have sufficient evidence to conclude that blocking is effective in reducing experimental error at $\alpha = 0.05$.

11.23 Summary Statistics: $\sum y_i = 1520.3$, $\sum y_i^2 = 110587$, $n = 21$,

$T_1 = 497.7$, $T_2 = 531.3$, $T_3 = 491.3$,

$B_1 = 211.1$, $B_2 = 202.7$, $B_3 = 233.1$,

$B_4 = 218.1$, $B_5 = 220.5$, $B_6 = 205.3$, $B_7 = 229.5$,

$$\text{TSS} = 110587.13 - \frac{(1520.3)^2}{21} = 524.6494,$$

$$\text{SST} = \frac{1}{6}\left((497.7)^2 + \cdots + (491.3)^2\right) - \frac{(1520.3)^2}{21} = 131.9010,$$

$$\text{SSB} = \frac{1}{3}\left((211.1)^2 + \cdots + (229.5)^2\right) - \frac{(1520.3)^2}{21} = 268.2894$$

ANOVA Table

Source	df	SS	MS	F
Treatment	2	131.9010	65.9505	6.3588
Block	6	268.2894	44.7149	4.3113
Error	12	124.4591	10.3716	
Total	20	524.6494		

Hypotheses: $H_0 : \mu_1 = \mu_2 = \mu_3$ H_a: At least two means differ

Test Statistic: F = 6.3588

Rejection Region: $F > F_{0.05}(2, 12) = 3.89$

Conclusion: Reject H_0 at $\alpha = 0.05$; i.e., there are significant differences in pressures required to separate the components among the three bonding agents.

11.25   Summary Statistics: $\sum y_i = 437$, $\sum y_i^2 = 18,169$, $n = 12$,

$T_1 = 110$, $T_2 = 109$, $T_3 = 218$,

$B_1 = 112$, $B_2 = 108$, $B_3 = 123$, $B_4 = 94$,

$$\text{TSS} = 18,169 - \frac{(437)^2}{12} = 2,254.9167,$$

$$\text{SST} = \frac{1}{4}\left((110)^2 + \cdots + (218)^2\right) - \frac{(437)^2}{12} = 1,962.1667,$$

$$\text{SSB} = \frac{1}{3}\left((112)^2 + \cdots + (94)^2\right) - \frac{(437)^2}{12} = 143.5833$$

ANOVA Table

Source	df	SS	MS	F
Treatment	2	1962.1667	981.0833	34.4626
Block	3	143.5833	47.8611	1.925
Error	6	149.1667	24.8611	1.925
Total	11	2254.9167		

a. Hypotheses: $H_0 : \mu_1 = \mu_2 = \mu_3$ H_a: At least two means differ

Test Statistic: $F = 39.46$

Rejection Region: $F > F_{0.05}(2, 6) = 5.14$

Conclusion: Reject H_0 at $\alpha = 0.05$; i.e., there are significant differences between the mean numbers of blades among the three stations.

b. Hypotheses: H_0: Block means are equal

H_a: At least two block means differ

Test Statistic: $F = 1.93$

Rejection Region: $F > F_{0.05}(3, 6) = 4.76$

Conclusion: Fail to reject H_0 at $\alpha = 0.05$; i.e., there are no significant differences in the mean number of blades among the four months.

11.27   Summary Statistics: $\sum y_i = 236.2$, $\sum y_i^2 = 3732.62$, $n = 15$,

$T_1 = 73.3$, $T_2 = 81.5$, $T_3 = 81.4$, $B_1 = 49.2$,

$B_2 = 46.8$, $B_3 = 46.3$, $B_4 = 48.8$, $B_5 = 45.1$,

$$\text{TSS} = 3,732.62 - \frac{(236.2)^2}{15} = 13.2573,$$

$$\text{SST} = \frac{1}{5}\left((73.3)^2 + \cdots + (81.4)^2\right) - \frac{(236.2)^2}{15} = 8.8573,$$

$$\text{SSB} = \frac{1}{3}\left((49.2)^2 + \cdots + (45.1)^2\right) - \frac{(236.2)^2}{15} = 3.977,$$

ANOVA Table

Source	df	SS	MS	F
Treatment	2	8.8573	4.4287	83.8239
Block	4	3.9773	0.9943	18.8203
Error	8	0.4227	0.0528	
Total	14	13.2573		

Hypotheses: $H_0 : \mu_1 = \mu_2 = \mu_3$ H_a: At least two means differ

Test Statistic: F = 83.82

Rejection Region: $F > F_{0.05}(2, 8) = 4.46$

Conclusion: Reject H_0 at $\alpha = 0.05$; i.e., there are significant differences in mean delivery times among the three carriers.

11.29   Summary Statistics: $\sum y_i = 6,598, \quad \sum y_i^2 = 3,632,768, \quad n = 12$

$T_1 = 1,653, \quad T_2 = 1,702, \quad T_3 = 1,634, \quad T_4 = 1,609,$

$B_1 = 2,265, \quad B_2 = 2,136, \quad B_3 = 2,197,$

$$\text{TSS} = 3,632,768 - \frac{(6,598)^2}{12} = 4,967.6690,$$

$$\text{SST} = \frac{1}{3}\left((1,653)^2 + \cdots + (1,609)^2\right) - \frac{(6,598)^2}{12} = 1,549.6680,$$

$$\text{SSB} = \frac{1}{4}((2,265)^2 + \cdots + (2,197)^2) - \frac{(6,598)^2}{12} = 2,082.1690$$

ANOVA Table

Source	df	SS	MS	F
Treatment	3	1549.6680	516.5560	2.3202
Block	2	2082.1690	1041.0845	4.6761
Error	6	1335.8320	222.6387	
Total	11	4967.6690		

a. Hypotheses: $H_0 : \mu_1 = \mu_2 = \mu_3 = \mu_4$ H_a: At least two means differ

Test Statistic: F = 2.32

Rejection Region: $F > F_{0.05}(3,6) = 4.76$

Conclusion: Fail to reject H_0 at $\alpha = 0.05$; i.e., there are not significant differences in mean temperature among the four treatments.

b. Hypotheses: H_0: Block means are equal H_a: At least two block means differ

Test Statistic: $F = 4.68$

Rejection Region: $F > F_{0.05}(2,6) = 5.14$

Conclusion: Fail to reject H_0 at $\alpha = 0.05$; i.e., there are not significant differences among the batch means.

11.31   Model: $Y = \beta_0 + \beta_1 x_2 + \beta_2 x_2 + \beta_3 x_3 + \beta_4 x_4 + \beta_5 x_5 + \epsilon$ where

$$x_1 = \begin{cases} 1 & \text{Block 1} \\ 0 & \text{otherwise} \end{cases} \qquad x_2 = \begin{cases} 1 & \text{Block 2} \\ 0 & \text{otherwise} \end{cases} \qquad x_3 = \begin{cases} 1 & \text{Block 3} \\ 0 & \text{otherwise} \end{cases}$$

$$x_4 = \begin{cases} 1 & \text{Treatment A} \\ 0 & \text{otherwise} \end{cases} \qquad x_5 = \begin{cases} 1 & \text{Treatment B} \\ 0 & \text{otherwise} \end{cases}$$

a. Hypotheses: Testing differences among treatment means is equivalent to testing

$H_0 : \beta_4 = \beta_5 = 0$ against $H_a : \beta_4 \neq 0$ or $\beta_5 \neq 0$

Test Statistic: (Uses SAS output: see the Appendix.) $F = \dfrac{(32 - 7.5)/2}{7.5/6} = 9.8$

Rejection Region: $F > F_{0.05}(2,6) = 5.14$

Conclusion: Reject H_0 at $\alpha = 0.05$; i.e., there are significant differences among the treatment means.

b. Hypotheses: Testing differences among blocks is equivalent to testing

$H_0 : \beta_1 = \beta_2 = \beta_3 = 0$ against $H_a : \beta_1 \neq 0$ or $\beta_2 \neq 0$ or $\beta_3 \neq 0$

Test Statistic: (Uses SAS output: see the Appendix.) $F = 1.25$

Rejection Region: $F > F_{0.05}(3,8) = 4.07$

Conclusion: Fail to reject H_0 at $\alpha = 0.05$; i.e., there are not significant differences among the blocks at $\alpha = 0.05$.

11.33 a.

$$
\begin{array}{c}
& & \text{Auto} \\
& & 1 & D_1 & D_2 & D_3 \\
\text{Driver} & 2 & D_3 & D_1 & D_2 \\
& 3 & D_2 & D_3 & D_1
\end{array}
$$

where D_i is engine design i, $i = 1, 2, 3$.

b. Model: $Y = \beta_0 + \beta_1 x_1 + \beta_2 x_2 + \beta_3 x_3 + \beta_4 x_4 + \beta_5 x_5 + \beta_6 x_6 + \epsilon$ where

$$
x_1 = \begin{cases} 1 & \text{Auto 1} \\ 0 & \text{otherwise} \end{cases} \quad
x_2 = \begin{cases} 1 & \text{Auto 2} \\ 0 & \text{otherwise} \end{cases} \quad
x_3 = \begin{cases} 1 & \text{Driver 1} \\ 0 & \text{otherwise} \end{cases}
$$

$$
x_4 = \begin{cases} 1 & \text{Driver 2} \\ 0 & \text{otherwise} \end{cases} \quad
x_5 = \begin{cases} 1 & \text{Design 1} \\ 0 & \text{otherwise} \end{cases} \quad
x_6 = \begin{cases} 1 & \text{Design 2} \\ 0 & \text{otherwise} \end{cases}
$$

c. $H_0 : \beta_5 = \beta_6 = 0 \qquad H_a : \beta_5 \neq 0 \text{ or } \beta_6 \neq 0$

11.35 $\dfrac{396.3}{6} - \dfrac{397.4}{6} \pm 1.812\sqrt{1.3919\left(\dfrac{1}{6} + \dfrac{1}{6}\right)} = (-1.418, 1.051)$

11.37 $t_{0.1/6} \approx 2.269 \qquad$ (degrees of freedom $= 22$)

Treatments 1 and 2: $\dfrac{13.42}{12} - \dfrac{13.44}{12} \pm 2.269\sqrt{0.0100\left(\dfrac{1}{12} + \dfrac{1}{12}\right)} = -0.0017\pm$

$0.0926 = (-0.0943, 0.0910)$. Since this interval includes zero, the difference between treatment 1 and treatment 2 is not significant at the 90% confidence level.

Treatments 1 and 3: $\dfrac{13.42}{12} - \dfrac{11.62}{12} \pm 0.0926 = (0.0574, 0.2426)$. Since this interval does not include zero, the difference between treatments 1 and 3 is significant at the 90% confidence level.

Treatments 2 and 3: $\dfrac{13.44}{12} - \dfrac{11.62}{12} \pm 0.0926 = (0.0591, 0.2443)$. Since this interval does not include zero, the difference between treatments 2 and 3 is significant at the 90% confidence level.

11.39 Summary Statistics: Let A_i = pressure i total and B_j = time j total.

$$\sum y_i = 44.9, \quad \sum y_i^2 = 168.79, \quad n = 12,$$

$$A_1 = 21.2, \quad A_2 = 23.7, \quad B_1 = 21.9, \quad B_2 = 23,$$

$$T_{11} = 10.4, \quad T_{12} = 11.5, \quad T_{21} = 10.8, \quad T_{22} = 12.2,$$

$$\text{TSS} = 168.79 - \frac{(44.9)^2}{12} = 0.7892.$$

$$\text{Let SST} = \frac{1}{3}\left((10.4)^2 + \cdots + (12.2)^2\right) - \frac{(44.9)^2}{12} = 0.6292.$$

$$\text{SS(A)} = \frac{1}{6}\left((21.2)^2 + \cdots + (23.7)^2\right) - \frac{(44.9)^2}{12} = 0.5208$$

$$\text{SS(B)} = \frac{1}{6}\left((21.9)^2 + \cdots + (23)^2\right) - \frac{(44.9)^2}{12} = 0.1008$$

$$\text{SS(A} \times \text{B)} = 0.6292 - 0.5208 - 0.1008 = 0.0075$$

$$\text{SSE} = 0.7892 - 0.6292 = 0.1600.$$

a. ANOVA Table

Source	df	SS	MS	F
Treatments	3	0.6292		
A	1	0.5208	0.5208	26.0417
B	1	0.1008	0.1008	5.0417
A × B	1	0.0075	0.0075	0.3750
Error	8	0.1600	0.0200	
Total	11	0.7892		

Comparing to $F_{0.05}(1, 8) = 5.32$, only the pressure effect (treatment A) is significant at $\alpha = 0.05$.

b. $\dfrac{21.9}{6} - \dfrac{23}{6} \pm 2.306\sqrt{(0.02)\left(\dfrac{1}{6} + \dfrac{1}{6}\right)} = (-0.3716, 0.00495)$

11.41 Summary Statistics: Let A_i = concentration of reactant i total and B_j = addition rate j total.

$$\sum y_i = 1,700, \quad \sum y_i^2 = 160,714.8, \quad n = 18,$$

$$A_1 = 548.8, \quad A_2 = 564.3, \quad A_3 = 586.9,$$

$$B_1 = 557.8, \quad B_2 = 567.5, \quad B_3 = 574.7,$$

$$T_{11} = 180.4, \quad T_{12} = 183.5, \quad T_{13} = 184.9,$$

$$T_{21} = 184.2, \quad T_{22} = 188.2, \quad T_{23} = 191.9,$$

$$T_{31} = 193.2, \quad T_{32} = 195.8, \quad T_{33} = 197.9.$$

a. ANOVA Table

Source	df	SS	MS	F
Treatments	8	148.0442		
A	2	122.3677	68.1838	458.8507
B	2	23.9743	11.9871	89.8981
A × B	2	1.7022	0.4256	3.1915
Error	9	1.2001	0.1333	
Total	17	149.2442		

Comparing to $F_{0.05}(4, 9) = 3.63$, the interaction of concentration of reactant and addition is not significant. Thus, we test for "main effects" and find both significant.

b. $\dfrac{188.2}{2} \pm 2.262\sqrt{0.1333\left(\dfrac{1}{2}\right)} = (93.516, 94.684)$

11.43 Summary Statistics: Let A_i = weight i total and B_j = sex j total.

$$\sum y_i = 2833.4, \quad \sum y_i^2 = 218357.6583, \quad n = 40,$$

$$A_1 = 1343, \quad A_2 = 1490.1, \quad B_1 = 1089, \quad B_2 = 1743.6,$$

$$T_{11} = 462.6, \quad T_{12} = 627.2, \quad T_{21} = 880.7, \quad T_{22} = 862.9,$$

$$\text{TSS} = 218,357.6583 - \frac{(2,833.4)^2}{40} = 17,653.7693,$$

$$\text{SST} = \frac{1}{10}\left((462.6)^2 + \cdots + (862.9)^2\right) - \frac{(2,833.4)^2}{40} = 12,056.861,$$

$$\text{SS(A)} = \frac{1}{20}\left((1,343.3)^2 + (1,490.1)^2\right) - \frac{(2,833.4)^2}{40} = 538.756,$$

$$\text{SS(B)} = \frac{1}{20}\left((1089.8)^2 + (1743.6)^2\right) - \frac{(2833.4)^2}{40} = 10686.361,$$

$$\text{SS(A×B)} = 12056.861 - 538.756 - 10686.361 = 831.744,$$

$$\text{SSE} = 9(14.23)^2 + 9(13.97)^2 + 9(8.32)^2 + 9(12.45)^2 = 5596.9083$$

a. ANOVA Table

Source	df	SS	MS	F
Treatments	3	12056.8610		
A	1	538.7560	538.7560	3.4653
B	1	10686.3610	10686.3610	68.7360
A×B	1	831.7440	831.7740	5.3499
Error	36	5596.9083	155.4697	
Total	39	17653.7693		

Comparing to $F_{0.05}(1, 36) = 4.12$, sex-by-weight interaction is significant at $\alpha = 0.05$. Thus, we do not make "main effects" tests.

b. The sex-by-weight interaction is significant at $\alpha = 0.05$. Hence, weight gain by females results in a large increase in pulling force while a corresponding weight gain by males does not give a similar increase in pulling force (in fact, a slight decline is exhibited).

c. $88.07 - 46.26 \pm 1.96\sqrt{155.4697\left(\dfrac{1}{10} + \dfrac{1}{10}\right)} = 41.81 \pm 10.9293$

$= (30.881, 52.739)$.

Since the interval does not include zero, there is not a significant difference between force exerted by light men and women.

d. $86.29 - 62.72 \pm 10.9293 = (12.641, 34.499)$. Since the interval does not include zero, there is not a significant difference between force exerted by heavy men and women.

11.45   Summary Statistics: Let $A_i =$ head type i total and $B_j =$ machine type j total.

$\sum y_i = 121, \sum y_i^2 = 671, n = 32,$

$A_1 = 33, A_2 = 27, A_3 = 32, A_4 = 29,$

$B_1 = 47, B_2 = 74,$

$T_{11} = 8, T_{12} = 25, T_{21} = 7, T_{22} = 20,$

$T_{31} = 27, T_{32} = 5, T_{41} = 5, T_{42} = 24$

a. ANOVA Table

Source	df	SS	MS	F
Treatments	7	165.7188		
A	3	2.8438	0.9479	0.4764
B	1	22.7812	22.7812	11.4502
A×B	3	140.0938	46.6979	23.4712
Error	24	47.7500	1.9896	
Total	31	213.4688		

Comparing to $F_{0.05}(3, 24) = 3.01$, we find that machine-by-head interaction is significant at $\alpha = 0.05$. Hence, we do not proceed with tests on "main effects."

b. To determine which head to recommend for machine A, we compute simultaneous confidence intervals for the six combinations. Let y_i denote the mean for machine A and head $i, i = 1, 2, 3, 4$. Then, using a simultaneous confidence coefficient of 0.90, we have, $t_{0.1/12} \approx 2.575$ (degrees of freedom = 24)

and $2.575 \sqrt{1.9896 \left(\dfrac{1}{4} + \dfrac{1}{4} \right)} = 2.57.$

Heads	Simultaneous Confidence Intervals
1 and 2	$2 - 1.75 \pm 2.57 = (-2.32, 2.82)$
1 and 3	$2 - 6.75 \pm 2.57 = (-7.32, -2.18)$
1 and 4	$2 - 1.25 \pm 2.57 = (-1.82, 3.32)$
2 and 3	$1.75 - 6.75 \pm 2.57 = (-7.57, -2.43)$
2 and 4	$1.75 - 1.25 \pm 2.57 = (-2.07, 3.07)$
3 and 4	$6.75 - 1.25 \pm 2.57 = (2.93, 8.07)$

Since only intervals comparing head 3 do not include zero, we see that means for heads 1, 2, and 4 are not significantly different. However, head 3 has a mean which is significantly different from 1, 2, and 4. Since head 3 has the largest mean strain resistance for machine A and is significantly different from the other head types, we recommend head 3.

11.47

    a. Completely Randomized Design.

    b. Summary Statistics: $\sum y_i = 117.9, \sum y_i^2 = 933.33$,

$$T_A = 40.8, T_B = 48.9, T_C = 28.2,$$

$$n_A = 5, n_B = 6, n_C = 4,$$

$$\text{TSS} = 933.33 - \frac{(117.9)^2}{6} = 6.636,$$

$$\text{SST} = \frac{(40.8)^2}{5} + \frac{(48.6)^2}{6} + \frac{(28.2)^2}{4} = 3.579$$

ANOVA Table

Source	df	SS	MS	F
Treatments	2	3.5790	1.7895	7.0245
Error	12	3.0570	0.2547	
Total	14	6.6360		

Hypotheses: $H_0 : \mu_A = \mu_B = \mu_C$ H_a: At least two means differ

Test Statistic: $F = 7.02$

Rejection Region: $F > F_{0.01}(2, 12) = 6.93$

Conclusion: Reject H_0 at $\alpha = 0.01$; i.e., the mean time to completion of the task significantly differs for the three methods at $\alpha = 0.01$.

    c. $\dfrac{48.9}{6} \pm 2.179 \sqrt{\dfrac{0.2547}{6}} = (7.701, 8.599)$

11.49   Summary Statistics: $\sum y_i = 230.3, \sum y_i^2 = 4431.17, n = 12$,

$$T_A = 76.5, T_B = 78, T_C = 75.8,$$

$$B_1 = 58.2, B_2 = 56.2, B_3 = 57.9, B_4 = 58,$$

$$\text{TSS} = 4431.17 - \frac{(230)^2}{12} = 11.3292,$$

$$\text{SST} = \frac{1}{4}\left((76.5)^2 + \cdots + (75.8)^2\right) - \frac{(230.2)^2}{12} = 0.6317,$$

$$\text{SSB} = \frac{1}{3}\left((58.2)^2 + \cdots + (58)^2\right) - \frac{(230.2)^2}{12} = 0.8558$$

ANOVA Table

Source	df	SS	MS	F
Treatments	2	0.6317	0.3158	0.1925
Blocks	3	0.8558	0.2853	0.1739
Error	6	9.8417	1.6403	
Total	11	11.3292		

Comparing the F-value of 0.1925 to $F_{0.05}(2, 6) = 5.14$, we find no significant differences in the mean mileage ratings among the three brands of gasoline at $\alpha = 0.05$.

c. Comparing the F-value of 0.1739 to $F_{0.05}(3, 6) = 4.76$, we find no significant difference in the mean mileage for the four models.

d. $\dfrac{78}{4} - \dfrac{75.8}{4} \pm 3.707\sqrt{1.6403\left(\dfrac{1}{4} + \dfrac{1}{4}\right)} = (-2.807, 3.907)$

e. $t_{0.1/6} \approx 2.749$ (degrees of freedom $= 6$) and $2.749\sqrt{1.6403\left(\dfrac{1}{4} + \dfrac{1}{4}\right)} = 2.4896$

Brands	Simultaneous Confidence Intervals
A and B	$\dfrac{76.5}{4} - \dfrac{78}{4} \pm 2.4896 = (-2.865, 2.115)$
A and C	$\dfrac{76.5}{4} - \dfrac{75.8}{4} \pm 2.4896 = (-2.315, 2.665)$
B and C	$\dfrac{78}{4} - \dfrac{75.8}{4} \pm 2.4896 = (-1.940, 3.040)$

11.51

a. Summary Statistics: $\sum y_i = 128, \sum y_i^2 = 948, n = 24$,

$A_1 = 24, A_2 = 32, A_3 = 26, A_4 = 49$,

$B_1 = 47, B_2 = 39, B_3 = 42$,

$T_{11} = 6, T_{12} = 11, T_{13} = 4, T_{21} = 9, T_{22} = 4, T_{23} = 19$,

$T_{31} = 17, T_{32} = 1, T_{33} = 8, T_{41} = 15, T_{42} = 23, T_{43} = 11$,

$$\text{TSS} = 948 - \frac{(128)^2}{24} = 265.3333,$$

$$\text{SST} = \frac{1}{2}\left((6)^2 + \cdots + (11)^2\right) - \frac{(128)^2}{24} = 247.3333,$$

$$\text{SS(A)} = \frac{1}{6}\left((21)^2 + \cdots + (49)^2\right) - \frac{(9128)^2}{24} = 74.3333,$$

$$\text{SS(B)} = \frac{1}{8}\left((47)^2 + \cdots + (42)^2\right) - \frac{(128)^2}{24} = 4.0833,$$

$$\text{SS(A}\times\text{B)} = 247.333 - 74.3333 - 4.0833 = 168.9167,$$

$$\text{SSE} = 265.3333 - 247.3333 = 18$$

ANOVA Table

Source	df	SS	MS	F
Treatments	11	247.3333		
A	3	74.3333	24.7778	16.5185
B	2	4.0833	2.0417	1.3611
A×B	6	168.9169	28.1528	18.7685
Error	12	18.0000	1.500	
Total	23	265.3333		

b. Comparing the F-value 18.77 to $F_{0.05}(6, 12) = 3.00$ indicates that there is a significant interaction between the factors at $\alpha = 0.05$.

c. $t_{0.1/18} \approx 2.998$ (degrees of freedom = 12), $2.998\sqrt{1.5\left(\frac{1}{6} + \frac{1}{6}\right)} = 2.1199$ and

$2.998\sqrt{1.5\left(\frac{1}{8} + \frac{1}{8}\right)} = 1.8359$. Let $A_i = \frac{A_i}{6}$ and $B_j = \frac{B_j}{8}$.

Simultaneous Confidence Intervals

$A_1 - A_2 \pm 2.1199 = (-3.959, 0.292)$

$A_1 - A_3 \pm 2.1199 = (-2.959, 1.292)$

$A_1 - A_4 \pm 2.1199 = (-6.792, 2.541)$

$A_2 - A_3 \pm 2.1199 = (-1.126, 3.126)$

$A_2 - A_4 \pm 2.1199 = (-4.959, -0.708)$

$A_3 - A_4 \pm 2.1199 = (-5.959, -1.708)$

$B_1 - B_2 \pm 1.8359 = (-0.840, 2.840)$

$B_1 - B_3 \pm 1.8359 = (-1.215, 2.465)$

$B_2 - B_3 \pm 1.8359 = (-2.215, 1.465)$

The intervals that contain zero indicate a nonsignificant difference. Hence, levels 1, 2, and 3 of factor A are not significantly different, whereas level 4 is significantly different from the others at $\alpha = 0.05$. Also, the levels of factor B are not significantly different. To select the treatment combination with the largest mean, we take level 4 of factor A and any level of factor B.

11.53

a. Completely Randomized Design.

b. Summary Statistics: $\sum y_i = 1311, \sum y_i^2 = 108587,$

$T_1 = 506, T_2 = 400, T_3 = 405,$

$n_1 = 6, \quad n_2 = n_3 = 5,$

$\text{TSS} = 108587 - \dfrac{(1311)^2}{16} = 1166.9374,$

$\text{SST} = \dfrac{(506)^2}{6} + \dfrac{(400)^2}{5} + \dfrac{(405)^2}{5} - \dfrac{(1311)^2}{16} = 57.6042$

ANOVA Table

Source	df	SS	MS	F
Treatments	2	57.6042	28.8021	0.3375
Error	13	1109.3333	85.3333	
Total	15	1166.9374		

c. Comparing the F-value 0.34 to $F_{0.05}(2, 13) = 3.81$, we see that there is not a significant difference in the mean productivities for the three lengths of workdays at $\alpha = 0.05$.

d. $\dfrac{405}{5} \pm 1.771 \sqrt{\dfrac{85.3333}{5}} = (73.684, 88.316)$

11.55 Summary Statistics: Let A_i = payment method i total and B_j be the scheduling type j total.

$$\sum y_i = 1192, \sum y_i^2 = 91198, n = 16,$$

$$A_1 = 520, A_2 = 672, B_1 = 558, B_2 = 634,$$

$$T_{11} = 240, T_{12} = 280, T_{21} = 318, T_{22} = 354,$$

$$\text{TSS} = 91190 - \frac{(1192)^2}{16} = 2394,$$

$$\text{SST} = \frac{1}{4}\left((240)^2 + \cdots + (354)^2\right) - \frac{(1192)^2}{16} = 1806,$$

$$\text{SS(A)} = \frac{1}{8}\left((520)^2 + (672)^2\right) - \frac{(1192)^2}{16} = 1444,$$

$$\text{SS(B)} = \frac{1}{8}\left((558)^2 + (634)^2\right) - \frac{(1192)^2}{16} = 361,$$

$$\text{SS(A}\times\text{B)} = 1806 - 1444 - 361 = 1,$$

$$\text{SSE} = 2394 - 1806 = 588$$

a. ANOVA Table

Source	df	SS	MS	F
Treatments	3	1806		
A	1	1444	1444	29.47
B	1	361	361	7.37
A×B	1	1	1	0.02
Error	12	588	49	
Total	15	2394		

b. Comparing the F-value 0.02 to $F_{0.05}(1, 12) = 4.75$, we find that the interaction is not significant at $\alpha = 0.05$.

c. $t_{0.05/4} \approx 2.597$ (degrees of freedom = 12) and $2560\sqrt{49\left(\frac{1}{4} + \frac{1}{4}\right)} = 12.6714$

$$A_1 - A_2 \pm 12.6714 = \frac{520}{4} - \frac{672}{4} \pm 12.6714 = (-50.671, -25.329)$$

$$B_1 - B_2 \pm 12.6714 = \frac{558}{4} - \frac{634}{4} \pm 12.6714 = (-31.671, -6.329)$$

The hourly and piece rate is significantly higher than the hourly rate and the worker-modified schedule is significantly higher than the 8−5 schedule. Thus, we recommend the hourly and piece rate and the worker-modified schedule.

11.57

a. Randomized Block Design.

b. Summary Statistics: $\sum y_i = 57.4, \sum y_i^2 = 222, n = 15,$

$T_1 = 10.3, T_2 = 11.2, T_3 = 12.1, T_4 = 12.2, T_5 = 12.6,$

$B_1 = 18.8, B_2 = 20.1, B_3 = 18.5,$

$$\text{TSS} = 222 - \frac{(57.4)^2}{15} = 2.3493,$$

$$\text{SST} = \frac{1}{3}\left((10.3)^2 + \cdots + (12.6)^2\right) - \frac{(57.4)^2}{15} = 1.7293,$$

$$\text{SSB} = \frac{1}{5}\left((18.8)^2 + (20.1)^2 + (18.5)^2\right) - \frac{(57.4)^2}{15} = 0.2893$$

ANOVA Table

Source	df	SS	MS	F
Treatments	4	1.7293	0.4323	10.4673
Blocks	2	0.2893	0.1447	3.5036
Error	8	0.3307	0.0413	
Total	14	2.3493		

Comparing the F-value 10.47 with $F_{0.05}(4, 8) = 3.84$ indicates that there are significant differences in the mean soil pH levels among the five concentrations of lime at $\alpha = 0.05$.

c. Comparing the F-value 3.50 with $F_{0.05}(2, 8) = 4.46$ indicates that there is not a significant difference in soil pH levels among the locations at $\alpha = 0.05$.

11.59   Summary Statistics: $\sum y_i = 450, \sum y_i^2 = 13876, n = 15,$

$T_1 = 138, T_2 = 179, T_3 = 133,$

$$\text{TSS} = 13876 - \frac{(450)^2}{15} = 376$$

$$\text{SST} = \frac{(138)^2}{5} + \frac{(179)^2}{5} + \frac{(133)^2}{5} - \frac{(450)^2}{15} = 254.8$$

ANOVA Table

Source	df	SS	MS	F
Treatments	2	254.8	127.4	12.61
Error	12	121.2	10.1	
Total	14	376		

Comparing the F-value 12.61 with $F_{0.05}(2, 12) = 3.89$, we see that there are significant differences among the three batches.

b. To compare the three batches, we use 90% simultaneous confidence intervals with $t_{0.01/6} \approx 2.403$ (degrees of freedom = 12) and $2.403 \sqrt{10.1 \left(\frac{1}{5} + \frac{1}{5} \right)} = 4.8300$.

Batches	Simultaneous Confidence Intervals
1 and 2	$\frac{138}{5} - \frac{179}{5} \pm 4.8300 = (-13.03, -3.37)$
1 and 3	$\frac{138}{5} - \frac{133}{5} \pm 4.8300 = (-3.83, 5.83)$
2 and 3	$\frac{179}{5} - \frac{133}{5} \pm 4.8300 = (4.37, 14.03)$

Since batch 2 is significantly different from batches 1 and 3, we select batch 2 to give the largest mean brightness.

11.61 ANOVA Table

Source	df	SS	MS	F	p
Treatments	7	2341400.0			
Diameter	1	530450.0	530450.0	19.57	0.0013
Thickness	2	1352744.4	676372.2	24.95	0.0001
Temp.	2	201036.1	100518.1	3.71	0.0624
RH	2	257169.4	128584.7	4.74	0.0356
Error	10	271061.1			
Total	17	2612461.1			

At the 0.05 significance level, the factor Temperature is the only one not significant.

11.63   For As/Cu:

Hypotheses: $H_0 : \mu_1 = \mu_2 = \mu_3 = \mu_4$ H_a: At least one μ_i is different

Test Statistic: $F = \dfrac{\sum_{i=1}^{k} n_i (\bar{y}_i - \bar{y})^2 / (k-1)}{S_p^2}$ where $S_p^2 = \dfrac{\sum_{i=1}^{k} (n_i - 1) S_i^2}{n - k}$

In this case,

$$S_p^2 = \frac{(7-1)(0.15)^2 + (11-1)(0.12)^2 + (31-1)(0.067)^2 + (5-1)(0.37)^2}{54 - 4}$$

$$= 0.0192$$

and

$$F = \frac{7(0.46 - 0.5456)^2 + 11(0.48 - 0.5456)^2 + 31(0.56 - 0.5456)^2 + 5(0.72 - 0.5456)^2}{(4-1)0.0192}$$

$$= 4.4641.$$

Rejection Region: $F > F_{0.05}(3, 50) = 2.79$

Conclusion: Reject H_0 at $\alpha = 0.05$; i.e., the mean mass ratio of arsenic to copper is significantly higher at the plume than at the other sites.

For Cd/Cu:

Hypotheses: $H_0 : \mu_1 = \mu_2 = \mu_3 = \mu_4$ H_a: At least one μ_i is different

Test Statistic:

$$S_p^2 = \frac{(10-2)(0.017)^2 + (11-1)(0.024)^2 + (31-1)(0.011)^2 + (5-1)(0.022)^2}{57 - 4}$$

$$= 0.0003$$

$$F = \frac{10(0.068 - 0.0757)^2 + 11(0.087 - 0.0757)^2}{(4-1)0.0003}$$

$$+ \frac{31(0.074 - 0.0757)^2 + 5(0.077 - 0.0757)^2}{(4-1)0.0003} = 2.3317$$

Rejection Region: $F > F_{0.05}(3, 53) = 2.7791$

Conclusion: Do not reject H_0 at $\alpha = 0.05$; i.e., there is no significant difference in the mean mass ratio of cadmium to copper at any of the four sites.

For Pb/Cu:

Hypotheses: $H_0 : \mu_1 = \mu_2 = \mu_3 = \mu_4$ $\qquad$ H_a: At least one μ_i is different

Test Statistic:

$$S_p^2 = \frac{(13-1)(0.16)^2 + (11-1)(0.17)^2 + (49-1)(0.07)^2 + (4-1)(0.23)^2}{77-4}$$

$$= 0.0136$$

$$F = \frac{13(1.03 - 0.8786)^2 + 11(0.94 - 0.8786)^2 + 49(0.82 - 0.8786)^2 + 4(0.90 - 0.8786)^2}{(4-1)0.0136}$$

$$= 12.4890$$

Rejection Region: $F > F_{0.05}(3, 73) = 2.7300$

Conclusion: Reject H_0 at $\alpha = 0.05$; i.e., there is a significant difference in the mean mass ratio of lead to copper at the Tucson Research ranch site from 8/84–10/84 and the Bisbee site.

For Sb/Cu:

Hypotheses: $H_0 : \mu_1 = \mu_2 = \mu_3 = \mu_4$ $\qquad$ H_a: At least one μ_i is different

Test Statistic:

$$S_p^2 = \frac{(3-1)(0.019)^2 + (7-1)(0.018)^2 + (11-1)(0.016)^2 + (5-1)(0.034)^2}{26-4}$$

$$= 0.0004$$

$$F = \frac{3(0.073 - 0.0821)^2 + 7(0.078 - 0.0821)^2 + 11(0.10 - 0.0821)^2 + 5(0.054 - 0.0821)^2}{(4-1)(0.0099)}$$

$$= 6.5315$$

Rejection Region: $F > F_{0.05}(3, 22) = 3.0491$

Conclusion: Reject H_0 at $\alpha = 0.05$; i.e., there is a significant difference in the mean mass ratio of antimony to copper at the plume site and the Tucson Research Ranch site (8/84–9/85).

For Zn/Cu:

Hypotheses: $H_0 : \mu_1 = \mu_2 = \mu_3 = \mu_4$ H_a: At least one μ_i is different

Test Statistic:

$$S_p^2 = \frac{(13-1)(0.31)^2 + (11-1)(1.1)^2 + (49-1)(1.6)^2 + (5-1)(0.67)^2}{78-4}$$

$$= 1.8639$$

$$F = \frac{13(1.64 - 7.0682)^2 + 11(4.2 - 7.0682)^2 + 49(9.7 - 7.0682)^2 + 5(1.7 - 7.0682)^2}{(4-1)1.8639}$$

$$= 171.1504$$

Rejection Region: $F > F_{0.05}(3, 74) = 2.7283$

Conclusion: Reject H_0 at $\alpha = 0.05$; i.e., there is a significant difference in the mean mass ratio of zinc to copper for the Tucson Research ranch site $(8/84-9/85)$ compared with the other sites.

11.65   Exercise for student.

11.67   (Uses SAS output: see the Appendix.)

$$\text{Let } X_1 = \begin{cases} 0 & \text{if not in Basin EE} \\ 1 & \text{if in Basin EE} \end{cases}$$

$$X_2 = \begin{cases} 0 & \text{if not in Basin F} \\ 1 & \text{if in Basin F} \end{cases}$$

$$X_3 = \begin{cases} 0 & \text{if not in Basin G} \\ 1 & \text{if in Basin G} \end{cases}$$

$$X_4 = \begin{cases} 0 & \text{if not in Basin M} \\ 1 & \text{if in Basin M} \end{cases}$$

11.69   Standardize all the means, and test for outliers. (Exercise for student.)

Appendix

1.1

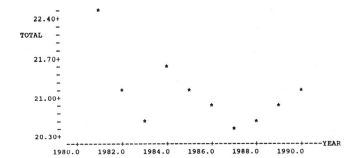

```
          -            *
   22.40+
          -
TOTAL     -
          -
          -
   21.70+
          -
          -                    *
          -
          -       *            *            *
   21.00+
          -                        *            *
          -
          -          *              *
          -                            *
   20.30+
        --+---------+---------+---------+---------+---------+----YEAR
       1980.0    1982.0    1984.0    1986.0    1988.0    1990.0
```

1.5

```
Stem-and-leaf of 1985      N  = 52            Stem-and-leaf of 1989      N  = 52
Leaf Unit = 0.10                              Leaf Unit = 0.10

    2    1 99                                      2    1 45
   10    2 00011111                                8    1 667777
   17    2 2233333                                14    1 889999
   24    2 4444455                                23    2 000111111
  (10)   2 6666666777                            (13)   2 2222333333333
   18    2 889                                    16    2 4444555
   15    3 0011                                    9    2 6
   11    3 2                                       8    2 88
   10    3 44455                                   6    3
    5    3 66                                      6    3 23333
    3    3 9                                       1    3 5
    2    4
    2    4 2
    1    4 4
```

```
                         ---------------
   1               ------I     +     I--------------     *  *
                         ---------------


                      ----------
   2          ---------I   +  I-------      *  *  *
                      ----------
              +---------+---------+---------+---------+---------+------C3
            1.20      1.80      2.40      3.00      3.60      4.20
```

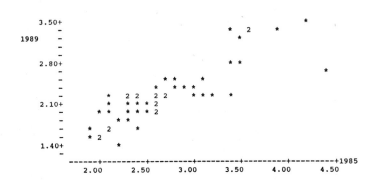

```
   3.50+
        -                                *  2       *          *
   1989 -                                   *
        -
        -
   2.80+                                *  *
        -                                               *
        -                       *  *     *
        -                     *    *  *  *
        -        *    2  2    2 2    *  *  *    *
   2.10+        *    *  *  *  2
        -     *  *    *  *  *  2
        -        *  *
        -     *    2     *
        -     *  2
   1.40+        *
        -
        ------+---------+---------+---------+---------+---------+1985
            2.00      2.50      3.00      3.50      4.00      4.50
```

1.9

```
Stem-and-leaf of CHANGE    N  = 18
Leaf Unit = 10

  (10)  -0 1000000000
    8    0 0000000
    1    0
    1    0
    1    0
    1    0
    1    1
    1    1
    1    1
    1    1
    1    1
    1    2 0
```

MTB > BOXPLOT C1

```
                    ----
                   -I +I-                                    o
                    ----
         ----+---------+---------+---------+---------+---------+--CHANGE
             0        40        80       120       160       200
```

MTB > HISTOGRAM C1

Histogram of CHANGE   N = 18

```
Midpoint    Count
    -20       1    *
      0      16    ****************
     20       0
     40       0
     60       0
     80       0
    100       0
    120       0
    140       0
    160       0
    180       0
    200       1    *
```

1.13

```
                                          ------
     1                                -----I +  I--
                                          ------

             -----------
     2       --I+        I-------
             -----------
         ------+---------+---------+---------+---------+---------+CHANGE
             0.80      1.60      2.40      3.20      4.00      4.80
```

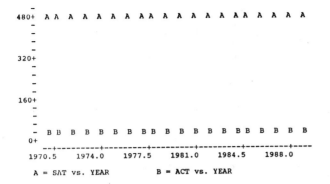

A = SAT vs. YEAR   B = ACT vs. YEAR

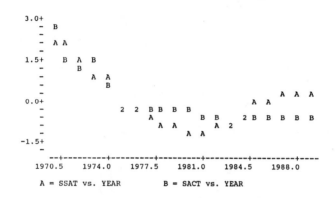

A = SSAT vs. YEAR   B = SACT vs. YEAR

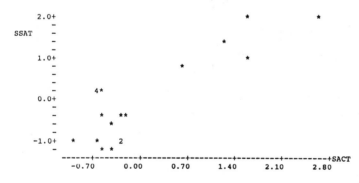

1.19

```
Stem-and-leaf of C1      N  = 51
Leaf Unit = 0.0010

    2    30  19
    4    31  68
   16    32  011366778899
   25    33  223336779
   (7)   34  0113334
   19    35  03334679
   11    36  1134689
    4    37
    4    38  88
    2    39  0
    1    40  6
```

MTB > BOXPLOT C1

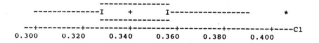

3.87

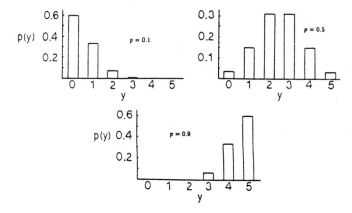

Q–Q Plot for Heights of Females aged 18–24

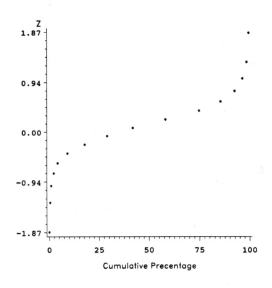

Residual Plot for Cumulative # Stamps Issued Model

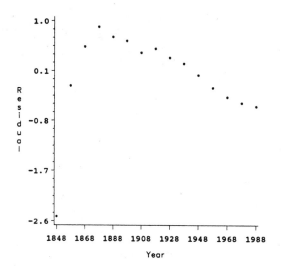

10.31

Model: MODEL1
Dependent Variable: TIME

Analysis of Variance

Source	DF	Sum of Squares	Mean Square	F Value	Prob>F
Model	2	1.27145	0.63573	150.617	0.0001
Error	7	0.02955	0.00422		
C Total	9	1.30100			

Root MSE	0.06497	R-square	0.9773	
Dep Mean	1.37000	Adj R-sq	0.9708	
C.V.	4.74218			

Parameter Estimates

Variable	DF	Parameter Estimate	Standard Error	T for H0: Parameter=0	Prob > |T|
INTERCEP	1	1.679390	0.04998155	33.600	0.0001
BRAND	1	0.444136	0.04110669	10.804	0.0001
MONTHS	1	-0.079322	0.00598078	-13.263	0.0001

Obs	Dep Var TIME	Predict Value	Std Err Predict	Lower95% Mean	Upper95% Mean	Lower95% Predict	Upper95% Predict
1	2.0000	1.9649	0.040	1.8703	2.0595	1.7845	2.1453
2	1.8000	1.8062	0.033	1.7283	1.8842	1.6340	1.9785
3	0.8000	0.7275	0.043	0.6269	0.8282	0.5439	0.9112
4	1.1000	1.1717	0.043	1.0689	1.2744	0.9869	1.3565
5	1.0000	1.0448	0.030	0.9740	1.1156	0.8757	1.2140
6	1.5000	1.5207	0.041	1.4242	1.6173	1.3393	1.7022
7	1.7000	1.6476	0.029	1.5784	1.7168	1.4791	1.8161
8	1.2000	1.2828	0.031	1.2095	1.3560	1.1126	1.4530
9	1.4000	1.4096	0.032	1.3330	1.4863	1.2379	1.5813
10	1.2000	1.1241	0.029	1.0554	1.1929	0.9558	1.2924
11	.	1.8062	0.033	1.7283	1.8842	1.6340	1.9785

Obs	Residual	Std Err Residual	Student Residual	-2-1-0 1 2	Cook's D
1	0.0351	0.051	0.686	| * |	0.096
2	-0.00624	0.056	-0.111	| |	0.001
3	0.0725	0.049	1.476	| ** |	0.546
4	-0.0717	0.048	-1.483	** | |	0.593
5	-0.0448	0.058	-0.777	* | |	0.054
6	-0.0207	0.051	-0.411	| |	0.037
7	0.0524	0.058	0.904	| * |	0.069
8	-0.0828	0.057	-1.450	** | |	0.206
9	-0.00963	0.056	-0.171	| |	0.003
10	0.0759	0.058	1.306	| ** |	0.142
11	.	.	.		.

Sum of Residuals		0
Sum of Squared Residuals		0.0295
Predicted Resid SS (Press)		0.0656

DEPENDENT VARIABLE: Y

SOURCE	DF	SUM OF SQUARES	MEAN SQUARE	F VALUE	PR > F
MODEL	5	17.58274143	3.51654829	100.41	0.0001
ERROR	34	1.19075607	0.03502224		R-SQUARE
CORRECTED TOTAL	39	18.77349750			0.936572

PARAMETER	ESTIMATE	T FOR HO: PARAMETER=0	PR>|T|	STD ERROR OF ESTIMATE
INTERCEPT	-9.91676311	-7.32	0.0001	1.35441340
X1	0.16680983	7.85	0.0001	0.02124474
X2	0.13759717	5.15	0.0001	0.02673395
X1*X1	-0.00110825	-9.45	0.0001	0.00011729
X2*X2	-0.00084327	-5.29	0.0001	0.00015942
X1*X2	0.00024109	1.67	0.1032	0.00014397

10.35

DEPENDENT VARIABLE: Y

SOURCE	DF	SUM OF SQUARES	MEAN SQUARE	F VALUE	PR > F
MODEL	1	14488.12209105	14448.12209105	124.37	0.0001
ERROR	7	815.43346450	116.49049493		R-SQUARE
CORRECTED TOTAL	8	15303.55555556			0.946716

PARAMETER	ESTIMATE	T FOR IIO: PARAMETER=0	PR>|T|	STD ERROR OF ESTIMATE
INTERCEPT	-93.12768073	-4.04	0.0049	23.04978592
X	0.44458154	11.15	0.0001	0.03986490

DEPENDENT VARIABLE: Y

SOURCE	DF	SUM OF SQUARES	MEAN SQUARE	F VALUE	PR > F
MODEL	2	15049.27786604	7524.63893302	177.55	0.0001
ERROR	6	254.27768952	42.37961492		R-SQUARE
CORRECTED TOTAL	8	15303.55555556			0.983384

PARAMETER	ESTIMATE	T FOR HO: PARAMETER=0	PR>|T|	STD ERROR OF ESTIMATE
INTERCEPT	204.46030828	2.46	0.0488	82.95426893
X	-0.63800270	-2.14	0.0764	0.29847798
X*X	0.00095925	3.64	0.0108	0.00026361

```
Model: MODEL1
Dependent Variable: REACTION

                        Analysis of Variance

                              Sum of          Mean
         Source       DF     Squares         Square     F Value     Prob>F

         Model         1    2117.50000    2117.50000        .           .
         Error         4            0             0
         C Total       5    2117.50000

              Root MSE        0.00000    R-square      1.0000
              Dep Mean       49.50000    Adj R-sq      1.0000
              C.V.            0.00000

                        Parameter Estimates

                      Parameter      Standard      T for H0:
         Variable  DF   Estimate        Error     Parameter=0    Prob > |T|

         INTERCEP   1  -7.10543E-15   0.00000000        .            .
         SPEED      1    1.100000     0.00000000        .            .



              Dep Var    Predict    Std Err   Lower95%   Upper95%   Lower95%   Upper95%
     Obs    REACTION      Value     Predict     Mean       Mean     Predict    Predict

      1     22.0000     22.0000      0.000    22.0000    22.0000    22.0000    22.0000
      2     33.0000     33.0000      0.000    33.0000    33.0000    33.0000    33.0000
      3     44.0000     44.0000      0.000    44.0000    44.0000    44.0000    44.0000
      4     55.0000     55.0000      0.000    55.0000    55.0000    55.0000    55.0000
      5     66.0000     66.0000      0.000    66.0000    66.0000    66.0000    66.0000
      6     77.0000     77.0000      0.000    77.0000    77.0000    77.0000    77.0000
      7        .        60.5000      0.000    60.5000    60.5000    60.5000    60.5000

                       Std Err    Student                          Cook's
     Obs    Residual   Residual   Residual    -2-1-0 1 2              D

      1    7.11E-15       .          .                               .
      2    7.11E-15       .          .                               .
      3    7.11E-15       .          .                               .
      4        0          .          .                               .
      5        0          .          .                               .
      6        0          .          .                               .
      7        .          .          .                               .

Sum of Residuals                    2.13E-14
Sum of Squared Residuals            0.0000
Predicted Resid SS (Press)          0.0000
```

Model: MODEL2
Dependent Variable: BRAKING

Analysis of Variance

Source	DF	Sum of Squares	Mean Square	F Value	Prob>F
Model	1	41382.91429	41382.91429	68.845	0.0012
Error	4	2404.41905	601.10476		
C Total	5	43787.33333			

Root MSE	24.51744	R-square	0.9451
Dep Mean	116.33333	Adj R-sq	0.9314
C.V.	21.07516		

Parameter Estimates

Variable	DF	Parameter Estimate	Standard Error	T for H0: Parameter=0	Prob > |T|
INTERCEP	1	-102.495238	28.20900935	-3.633	0.0221
SPEED	1	4.862857	0.58607886	8.297	0.0012

Obs	Dep Var BRAKING	Predict Value	Std Err Predict	Lower95% Mean	Upper95% Mean	Lower95% Predict	Upper95% Predict
1	20.0000	-5.2381	17.744	-54.5038	44.0276	-89.2660	78.7898
2	40.0000	43.3905	13.322	6.4039	80.3771	-34.0794	120.9
3	72.0000	92.0190	10.429	63.0629	121.0	18.0458	166.0
4	118.0	140.6	10.429	111.7	169.6	66.6744	214.6
5	182.0	189.3	13.322	152.3	226.3	111.8	266.7
6	266.0	237.9	17.744	188.6	287.2	153.9	321.9
7	.	165.0	11.599	132.8	197.2	89.6584	240.3

Obs	Residual	Std Err Residual	Student Residual	-2-1-0 1 2	Cook's D
1	25.2381	16.919	1.492	| ** |	1.224
2	-3.3905	20.582	-0.165	| |	0.006
3	-20.0190	22.189	-0.902	| * |	0.090
4	-22.6476	22.189	-1.021	| ** |	0.115
5	-7.2762	20.582	-0.354	| |	0.026
6	28.0952	16.919	1.661	| *** |	1.517
7	.	.	.	| |	.

Sum of Residuals	0
Sum of Squared Residuals	2404.4190
Predicted Resid SS (Press)	7781.7269

Model: MODEL3
Dependent Variable: TOTAL

Analysis of Variance

Source	DF	Sum of Squares	Mean Square	F Value	Prob>F
Model	1	62222.41429	62222.41429	103.513	0.0005
Error	4	2404.41905	601.10476		
C Total	5	64626.83333			

Root MSE	24.51744	R-square	0.9628	
Dep Mean	165.83333	Adj R-sq	0.9535	
C.V.	14.78438			

Parameter Estimates

Variable	DF	Parameter Estimate	Standard Error	T for H0: Parameter=0	Prob > |T|
INTERCEP	1	-102.495238	28.20900935	-3.633	0.0221
SPEED	1	5.962857	0.58607886	10.174	0.0005

Obs	Dep Var TOTAL	Predict Value	Std Err Predict	Lower95% Mean	Upper95% Mean	Lower95% Predict	Upper95% Predict
1	42.0000	16.7619	17.744	-32.5038	66.0276	-67.2660	100.8
2	73.0000	76.3905	13.322	39.4039	113.4	-1.0794	153.9
3	116.0	136.0	10.429	107.1	165.0	62.0458	210.0
4	173.0	195.6	10.429	166.7	224.6	121.7	269.6
5	248.0	255.3	13.322	218.3	292.3	177.8	332.7
6	343.0	314.9	17.744	265.6	364.2	230.9	398.9
7	.	225.5	11.599	193.3	257.7	150.2	300.8

Obs	Residual	Std Err Residual	Student Residual	-2-1-0 1 2	Cook's D
1	25.2381	16.919	1.492	|**	1.224
2	-3.3905	20.582	-0.165	|	0.006
3	-20.0190	22.189	-0.902	*|	0.090
4	-22.6476	22.189	-1.021	**|	0.115
5	-7.2762	20.582	-0.354	|	0.026
6	28.0952	16.919	1.661	|***	1.517
7	.	.	.		.

Sum of Residuals		0
Sum of Squared Residuals		2404.4190
Predicted Resid SS (Press)		7781.7269

```
------------------------------ COMPOUND= ----------------------------------
                                   BaP
Model: MODEL1
Dependent Variable: PERCENT

                        Analysis of Variance

                          Sum of        Mean
      Source       DF     Squares       Square      F Value      Prob>F

      Model         1   529.92701     529.92701      12.464       0.0242
      Error         4   170.07299      42.51825
      C Total       5   700.00000

         Root MSE      6.52060     R-square      0.7570
         Dep Mean     11.00000     Adj R-sq      0.6963
         C.V.         59.27820

                        Parameter Estimates

                     Parameter     Standard     T for H0:
      Variable  DF    Estimate       Error     Parameter=0    Prob > |T|

      INTERCEP   1   -25.590198   10.70080771     -2.391        0.0751
      MEANRF     1    68.821689   19.49418050      3.530        0.0242



------------------------------ COMPOUND= ----------------------------------
                                   BaA
Model: MODEL1
Dependent Variable: PERCENT

                        Analysis of Variance

                          Sum of        Mean
      Source       DF     Squares       Square      F Value      Prob>F

      Model         1   604.27082     604.27082      25.249       0.0074
      Error         4    95.72918      23.93230
      C Total       5   700.00000

         Root MSE      4.89206     R-square      0.8632
         Dep Mean     11.00000     Adj R-sq      0.8291
         C.V.         44.47331

                        Parameter Estimates

                     Parameter     Standard     T for H0:
      Variable  DF    Estimate       Error     Parameter=0    Prob > |T|

      INTERCEP   1   -20.633835    6.60467035     -3.124        0.0354
      MEANRF     1    77.470617   15.41747987      5.025        0.0074
```

```
-------------------------------- COMPOUND= --------------------------------
                                    Phe
Model: MODEL1
Dependent Variable: PERCENT

                          Analysis of Variance

                              Sum of        Mean
            Source       DF   Squares       Square      F Value     Prob>F

            Model        1   616.78344    616.78344     29.647      0.0055
            Error        4    83.21656     20.80414
            C Total      5   700.00000

                Root MSE       4.56116     R-square      0.8811
                Dep Mean      11.00000     Adj R-sq      0.8514
                C.V.          41.46505

                          Parameter Estimates

                      Parameter      Standard     T for H0:
            Variable  DF  Estimate      Error    Parameter=0    Prob > |T|

            INTERCEP   1  -16.097181   5.31355735    -3.029        0.0388
            MEANRF     1   86.022795  15.79872622     5.445        0.0055
```

DEPENDENT VARIABLE: Y

SOURCE	DF	SUM OF SQUARES	MEAN SQUARE	F VALUE	PR > F
MODEL	5	39.5000000	7.90000000	6.32	0.0220
ERROR	6	7.50000000	1.25000000		R-SQUARE
CORRECTED TOTAL	11	47.0000000			0.840426

PARAMETER	ESTIMATE	T FOR HO: PARAMETER=0	PR>|T|	STD ERROR OF ESTIMATE
INTERCEPT	1.75000000	2.21	0.0688	0.79056942
X1	0.33333333	0.37	0.7275	0.91287093
X2	2.33333333	2.56	0.0431	0.91287093
X3	-0.66666667	-0.73	0.4927	0.91287093
X4	0.50000000	0.63	0.5504	0.79056942
X5	3.25000000	4.11	0.0063	0.79056942

DEPENDENT VARIABLE: Y

SOURCE	DF	SUM OF SQUARES	MEAN SQUARE	F VALUE	PR>F
MODEL	3	15.00000000	5.00000000	1.25	0.3544
ERROR	8	32.00000000	4.00000000		R-SQUARE
CORRECTED TOTAL	11	47.00000000			0.319149

PARAMETER	ESTIMATE	T FOR HO: PARAMETER=0	PR>|T|	STD ERROR OF ESTIMATE
INTERCEPT	3.00000000	2.60	0.0317	1.15470054
X1	0.33333333	0.20	0.8434	1.63299316
X2	2.66666667	1.43	0.1909	1.63299316
X3	-0.66666667	-0.41	0.6938	1.63299316

General Linear Models Procedure

Dependent Variable: MEANSAL

Source	DF	Sum of Squares	Mean Square	F Value	Pr > F
Model	5	0.34548908	0.06909782	3.85	0.0292
Error	11	0.19746598	0.01795145		
Corrected Total	16	0.54295506			

R-Square	C.V.	Root MSE	MEANSAL Mean
0.636312	71.73895	0.1339830	0.1867647

Source	DF	Type I SS	Mean Square	F Value	Pr > F
BASIN	4	0.31355289	0.07838822	4.37	0.0234
DEPTH	1	0.03193619	0.03193619	1.78	0.2092

Source	DF	Type III SS	Mean Square	F Value	Pr > F
BASIN	4	0.28291925	0.07072981	3.94	0.0319
DEPTH	1	0.03193619	0.03193619	1.78	0.2092

Parameter		Estimate	T for H0: Parameter=0	Pr > |T|	Std Error of Estimate
INTERCEPT		-.0067195779 B	-0.07	0.9419	0.09013902
BASIN	EE	0.1031651350 B	0.94	0.3697	0.11030402
	F	0.0544897125 B	0.53	0.6064	0.10274324
	G	0.3910089159 B	3.53	0.0047	0.11063215
	M	0.0861999552 B	0.82	0.4311	0.10546747
	MM	0.0000000000 B	.	.	.
DEPTH		0.0061946716	1.33	0.2092	0.00464437

NOTE: The X'X matrix has been found to be singular and a generalized inverse was used to solve the normal equations. Estimates followed by the letter 'B' are biased, and are not unique estimators of the parameters.